# A New Earth Reinforcement Method
# Using Soilbags

BALKEMA – Proceedings and Monographs
in Engineering, Water and Earth Sciences

# A New Earth Reinforcement Method Using Soilbags

Hajime Matsuoka

*Nagoya Institute of Technology, Nagoya, Japan*

Sihong Liu

*Hohai University, Nanjing, People's Republic of China*

## CRC Press
Taylor & Francis Group
Boca Raton London New York

CRC Press is an imprint of the
Taylor & Francis Group, an **informa** business

A BALKEMA BOOK

Typeset in Times New Roman by
Integra Software Services Pvt. Ltd, Pondicherry, India

Published by:    Taylor & Francis / Balkema
                 P.O. Box 447, 2300 AK Leiden, The Netherlands
                 e-mail: Pub.NL@tandf.co.uk
                 www.balkema.nl, www.tandf.co.uk, www.crcpress.com

*British Library Cataloguing in Publication Data*
A catalogue record for this book is available from the British Library

*Library of Congress Cataloging in Publication Data*
A catalog record for this book has been requested

ISBN 13: 978-0-415-38354-7 (hbk)

# Contents

# Preface

The first author has been studying the mechanics of granular materials for many years and has devoted his research to the investigation of soil particle movement under external loads. This microscopic study provides valuable insight into the fundamental properties of granular soils, and has led to the well-recognized three-dimensional failure criterion for soils, known as the SMP criterion.

About a decade ago, we commenced laboratory tests on the assembly of aluminum rods to explore the characteristics of granular soils in the two-dimensional frame. The study shed light on the re-evaluation of soilbags (wrapping soil in bags). In fact, soilbags have been commonly used in temporary structures rather than earth reinforcement because of the deterioration of the bags after long exposure to sunlight. Nevertheless, *in situ* experiments in our laboratory demonstrated the amazing bearing capacity of soilbags. Analogously, we developed a novel and effective earth reinforcement method using soilbags, so that the bearing capacity of the soft foundation can be greatly improved. The bearing load per unit of quality-controlled soilbags can reach 10–20% of concrete. Without direct exposure to sunlight (ultraviolet rays), soilbags can be used as a semi-permanent material (>50 years) in engineering applications. This book introduces the soilbag earth reinforcement method with some detailed documentation of its vast geotechnical applications in Japan. We hope that this method could be widely adopted in earth reinforcement and civil engineering construction in many other countries, especially in developing countries, as an effective and economical alternative to earth reinforcement.

We are most appreciative of the contributions made by Dr. H. Yamamoto (Hiroshima University) who first proposed the construction of soilbag piles and elucidated the high damping properties of soilbags during his study at Nagoya Institute of Technology (NIT). We also sincerely thank Dr. K. Yamaguchi (San-ei House Co., Ltd.) for his active promotion of the earth reinforcement method of soilbags, Dr. M. Tateyama (JR General Research Institute) and Mr. T. Kachi (JR Tokai Corporation) for their much-valued assistance in applying the soilbags to the reinforcement of railway foundations. We acknowledge with deep appreciation Prof. Y. Chen, former visiting researcher at NIT from Suzhou Institute of Science and Technology, People's Republic of China, who provided great help in the theoretical study on the mechanical behaviors of soilbags, and Dr. Y. Kusano, former researcher at Technical Institute of Civil Engineering, the Tokyo Metropolitan, who actively applied this reinforcement method in Miyake Island.

We would also like to thank the people at Solpack Associate, especially Dr. I. Tanahashi (the former President), Dr. S. Chida (President) and H. Satoh (Techno. Sol.). Acknowledgments are also made to the following individuals: H. Takamori (WASC, Co. Ltd. Corporation), S. Maeda (S. Kato Co. Ltd.) and Y. Yokota (Maeda Kosen Co. Ltd.), H. Teramoto (Fuji Engineering Co. Ltd.), H. Andoh (Taiyu Construction Co. Ltd.), Y. Kitamura (Civil Engineering Bureau of Nagoya City) and M. Futaki (Financial Better Living).

We would like to express our gratitude to Dr. M. Masuda, Associate Professor at NIT for her collaboration in the study of natural vegetation planted in soilbags, to Prof. H. Umehara, Associate Professor, T. Uehara and Mr. H. Hirahara at the concrete research laboratory of NIT for their help in the unconfined compression tests on soilbags. Thanks are also due to Prof. T. Suzuki of Kitami Institute of Technology for his collaboration in the study of frost heave prevention by the use of soilbags.

Many students have been particularly helpful in conducting the laboratory experiments: N. Takagi, M. Nishii, S. Iwai, S. Okuda, T. Nishimura, K. Miyamoto, T. Ono, T. Takizawa, K. Itoh, T. Ueda, T. Saeki, H. Kodama, Z. Nakamura, Y. Izuka, T. Oka, J. Nakamura, K. Yamaji, T. Hasebe, R. Shimao, M. Yonetani, K. Fujita, T. Yamada, D. Muramatsu, T. Inoue, S. Hattori, Y. Asano and T. Feng.

Very special thanks go to Dr. H. Chen (Golder Associates, Ltd.) for the thorough review of the English language and text of the book.

Hajime Matsuoka
*Nagoya Institute of Technology, Nagoya, Japan*
Sihong Liu
*Hohai University, Nanjing, P.R. China*
*March 2005*

# Chapter 1

# Why do we study soilbags now?

Soilbags, "Donow" in Japanese, are commonly used for embankment raising at times of inundation and as temporary structures during reconstruction after disasters. Soilbags have rarely been applied to earth reinforcement because of the deterioration of bags after long exposure to sunlight, especially the polyethylene (PE) bags that are extremely vulnerable to ultraviolet rays. Nevertheless, there are advantages to using quality-controlled soilbags (wrapping soils in a particular type of bag) as earth reinforcement (Table 1.1). This is because, when a soilbag undergoes external force or building load, tensile force occurs along the bag, which in turn enhances the bearing capacity of the soilbag. Analogously, external forces including the self-weight of buildings, will strengthen the foundations reinforced by soilbags. It is interesting to see that soilbags have the ability to convert the external force which is the "enemy" of the foundations into the "friend" of the foundations using the action of tensile forces along the bags. Inspired by these unique characteristics, soilbags may be applied to soft foundation reinforcement. The bearing capacity of the foundations will be greatly improved, although a small amount of settlement still exists in the reinforced foundations. Wrapping soils in a particular type of bag is more effective and reliable than the commonly used horizontal sheet earth reinforcement. The advantages of the quality-controlled soilbags may be summarized as follows:

- Bags are cheap and easy to acquire.
- Soilbags have almost the same unit weight as foundation soils.
- The materials inside soilbags can be various construction wastes, such as crushed concrete, asphalt and tile wastes. Soilbags thus contribute greatly to the recycling of waste materials.
- No special construction equipment is required. Soilbags can be constructed solely by human labor.
- Earth reinforcement using soilbags is environmentally friendly because cement and chemical agents are avoided.
- Less noise and vibration are produced during construction, in comparison to the pile-driving method that is commonly used in soft/weak foundation reinforcement.

*Table 1.1* Characteristics of soilbags

- It makes sense to wrap discrete soil particles.
- The bearing capacity of foundations may be increased significantly (by 5–10 times).
- Bags are cheap and easy to acquire.
- Earth reinforcement using soilbags is environmentally friendly, because cement and chemical agents are avoided.
- Less noise and vibration are produced during construction, in comparison to the pile-driving method that is commonly used in soft foundation reinforcement.
- No special construction equipment is required. Soilbags can be constructed solely by human labor.
- Soilbags have almost the same unit weight as foundation soils.
- The materials inside soilbags can be various construction wastes, such as crushed concrete, asphalt and tile wastes. Soilbags thus contribute greatly to the recycling of waste materials.

*Additional effects*
- The compressive force of soilbags is amazingly high (200–300 kN).
- Traffic- or machine-induced vibrations may be reduced effectively.
- Frost heave may be reduced if soilbags are filled with coarse granular materials.
- Waterlogged soft ground can also be reinforced using soilbags.

The aforementioned advantages of soilbags have been predicted through our laboratory and *in situ* experiments. More advantages have been demonstrated in engineering practices, including the amazing bearing capacity, the reduction of traffic-induced vibration, the resistance to earthquakes, the prevention of frost heave and the reinforcement on waterlogged soft ground. For example, an ordinary PE bag filled with crushed stones or sand (approx. 40 cm × 40 cm × 10 cm) can withstand a load of up to 230–280 kN. If the high strength polyester (PET) bags are used, the bearing load may be increased to 540–640 kN. Such high values are higher than expected, up to ten times higher than the predictions made by experienced workers.

Soils are essentially frictional materials. Wrapping frictional earth materials in a bag is an innovative idea. In comparison with steel and concrete, the materials used in soilbag construction are flexible and environmentally friendly. Therefore, quality-controlled soilbags may be widely used in earth reinforcement (Matsuoka and Liu, 2003).

# Chapter 2

# How to achieve earth reinforcement with soilbags (Solpack method)?

Compared with steel and concrete, soil is essentially an assembly of distinct particles and usually breaks into pieces at failure. Resistance forces between soil particles are basically frictional forces. Wrapping soil in a bag may increase the normal force $N$ between soil particles, which in turn raises the frictional force $F$ between soil particles ($F = \mu N$, where $\mu$ is the friction coefficient). Therefore, wrapping soils in bags is not only a simple and effective method of earth reinforcement; it also turns the negative effect of external force on a foundation into a positive effect.

Figure 2.1 schematically shows the setup of the bearing capacity test. The foundation is modeled by 5 cm long aluminum rods with two different diameters of 1.6 and 3 mm. The mixed aluminum rod assembly has a weight ratio of 3:2, a void ratio $e$ of 0.23, and a dry density $\gamma$ of 21.6 kN/m$^3$. Due to the following advantages, the aluminum rods accurately simulate granular materials in a two-dimensional manner:

- The specific gravity $G_s$ of the aluminum rods is 2.69, close to that of the real soil particles (about 2.65).
- The assembly of aluminum rods can support its own weight without any additional support on the front/back sides in which no frictional resistances are produced.
- The movement of the particles (aluminum rods) can be easily observed and traced by drawing lines on the front surface.

A series of bearing capacity tests were conducted on the assembly of aluminum rods to develop a new foundation reinforcement method. Part of the aluminum rods was wrapped in a piece of paper. The mechanism of the reinforcement was investigated through the observation of the movement of the aluminum rods.

In order to improve the bearing capacity by 10 times, we are going to develop a new reinforcement method that is different from the traditional ones. The traditional methods usually place the reinforcing materials (such as geotextiles, nets, mattresses, strips, etc.) horizontally into the foundation (e.g. Binquest and Lee, 1975a,b; Huang and Tatsuoka, 1990; Hirao *et al.*, 1997). Nevertheless, we still start our tests in the traditional way. A piece of 30 cm long paper is placed horizontally into the

*Figure 2.1* Setup of model footing bearing capacity test.

aluminum rods at a depth of 3 cm beneath the footing, as shown in Figure 2.2. The allowable tensile force of the paper is 33–41 N per cm, and the paper is 64 g per square meter. The test only shows a small increase in bearing capacity, which is due to the slip of the aluminum rods above the horizontal paper and the deformation of the paper. We recall the major principal stress distributions of strip footing foundation in the theory of elasticity. As shown in Figure 2.3, the minor principal strain $\varepsilon_3$ or the major tensile strain takes place almost along the direction of the minor principal stress $\sigma_3$. It is expected that placing flexible reinforced materials along the direction of the minor principal strain $\varepsilon_3$ would be the most effective way to reinforce foundations. This is because the extendable reinforced material should be placed in the most extendable direction where the material may

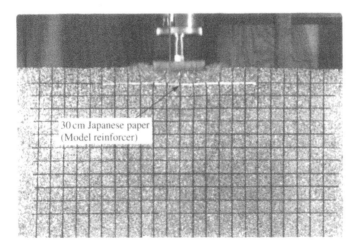

30 cm Japanese paper
(Model reinforcer)

*Figure 2.2* Reinforcement with a horizontal piece of paper (30 cm long and 3 cm deep).

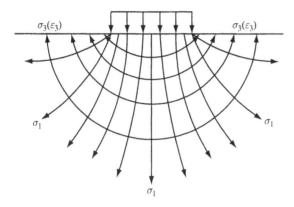

Figure 2.3 Orientations of principal stresses under a footing from an elastic calculation.

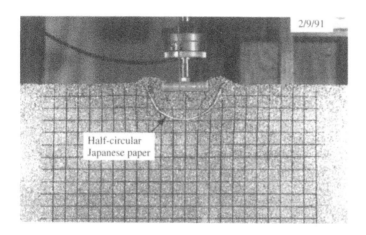

Figure 2.4 Reinforcement with a semi-circular piece of paper open in two sides of the footing.

develop maximum tensile deformation. Therefore, a piece of paper is placed along the semi-circular direction of $\varepsilon_3$ in the aluminum rods, as shown in Figure 2.4. As expected, the bearing capacity of the aluminum rod assembly is increased due to the frictional resistance between the flexible paper and the rods. However, the increment is not enough because the rods move upwards from the two sides of the footing (Figure 2.4). To overcome this problem, we extend both sides of the flexible paper to the bottom of the footing, that is, part of the rods is wrapped in the paper (Figure 2.5). As a result, the bearing capacity of the rods, that is in the simulated foundation, increases dramatically.

The mechanism of reinforcement is to be experimentally investigated in the following three situations. The first test setup is shown in Figure 2.6, where part of the aluminum rod beneath the footing is wrapped in a semi-circular piece of

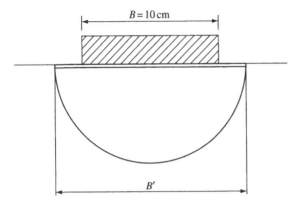

*Figure 2.5* Wrapping a part of the foundation beneath a footing, where B is the width of the footing and B′ is the initial width of the part wrapped by a reinforcer.

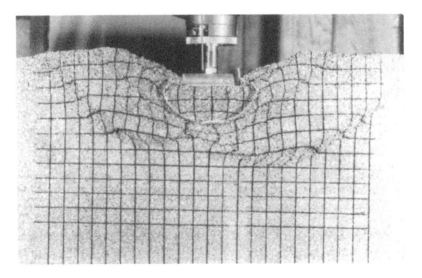

*Figure 2.6* Sliding failure in the aluminum rod foundation reinforced with a piece of the semi-circular paper (B = 10 cm, B′ = 15 cm).

paper (footing width $B = 10$ cm, initial width of the wrapped material $B' = 15$ cm). It can be seen that a large slip line is generated under the wrapped ground. This is because, when the external load is applied on the footing, the wrapped material behaves as part of the footing. The second test setup is shown in Figure 2.7, where the wrapped material is replaced with a semi-circular wood block. In the third test (Figure 2.8), the wrapped material is stabilized with gum-tapes on both the front and the back surfaces of the aluminum rods. We supposed that the bearing capacity of the foundation would be the largest in Cases 2 and 3. However, beyond our expectation, the largest bearing capacity we measured was Case 1 (Figure 2.9),

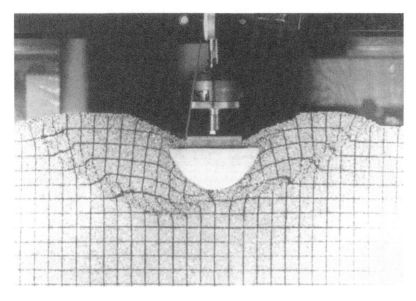

Figure 2.7 Sliding failure in the aluminum rod foundation reinforced with one semi-circular wood block ($B = 10$ cm, $B' = 15$ cm).

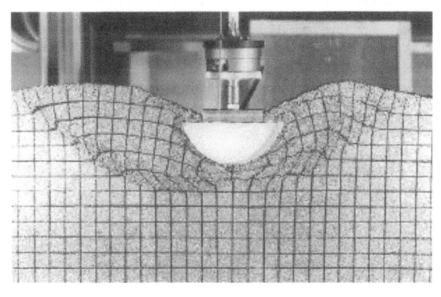

Figure 2.8 Sliding failure in the aluminum rod foundation with a semi-circular part pasted and fixed by gum-tape ($B = 10$ cm, $B' = 15$ cm).

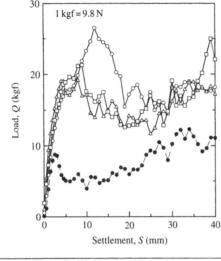

1 kgf = 9.8 N

—○— Wrapping a semi-circular piece of ground with a piece of paper
—□— Pasting and fixing a semi-circular part of ground with tapes
—△— Reinforcing the ground with a semi-circular wood block
—●— No reinforcement

*Figure 2.9* Relationship between load Q and settlement S of the footing under different reinforcements for a semi-circular part of the aluminum rod foundation ($B = 10$ cm, $B' = 15$ cm).

that is the part of the ground wrapped with a piece of paper. Why does Case 1 have the largest bearing capacity? Let us have a close look at the wrapped material subjected to external forces (Figure 2.10). The width of the semi-circular-shaped material is increased up to several centimeters due to lateral expansion. Moreover, the wrapped material seems to have been solidified, as if it were integrated with

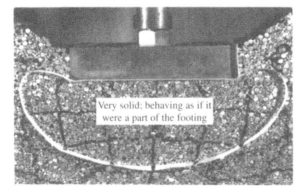

Very solid; behaving as if it were a part of the footing

*Figure 2.10* Enlarged picture of the semi-circular wrapped part subjected to an external force.

the footing. The external force applied on the footing induces tensile force along the paper as a result of the paper extension (cf. Figure 3.2). The wrapped particles (aluminum rods) inside the paper are restrained, leading to an increase in the effective stress $\sigma'$. Subsequently, the shear strength $\tau_f$ of the wrapped particles increases ($\tau_f = \sigma' \tan\phi'$). As part of the footing, the solidified material thus increases greatly the bearing capacity of the foundation. Essentially, this hypothesis is analogous to the increase in normal force $N$ leading to the increase in frictional force $F(=\mu N)$. It is interesting that the hypothesis involves a "reversal idea" of using external force to reinforce foundations, which was usually the "enemy" of foundations (Matsuoka et al., 1992). Figure 2.11 shows the numerical simulation results by the Distinct Element Method (DEM), in which Figure 2.11(a) is the displacement distribution of particles relative to the footing (settlements of the footing = 18.0−22.5 mm), and Figure 2.11(b) describes the interparticle contact forces at the maximum bearing capacity of the reinforced ground (Yamamoto and Matsuoka, 1995). It can be seen from Figure 2.11(a) that the wrapped particles do not have relative movement to the footing. Also, a wedge-shaped zone is formed beneath the wrapped body. It seems as if the wrapped particles have integrated with the footing to form a wider and deeper one. Figure 2.11(b) illustrates that the interparticle contact forces inside the wrapped body are considerably larger

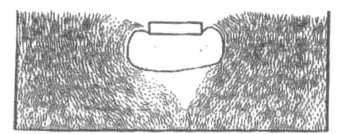

(a) Vectors of particle displacements relative to the footing when the footing settles from 18.0 mm to 22.5 mm

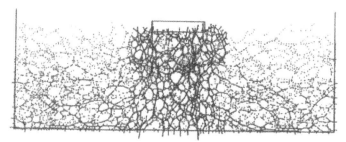

(b) Transmission network of interparticle forces at the peak bearing load

*Figure 2.11* Numerically simulated results by DEM, illustrating the mechanism of the increase in bearing load of the ground reinforced using soilbags (after Yamamoto and Matsuoka, 1995).

than those outside, that is, the effective stresses inside the wrapped body are much larger than those outside.

In order to investigate the behavior of the wrapped body filled with loose materials, part of the aluminum rod is removed from the semi-circular body so as to increase its initial void ratio from 0.23 to 0.28 and 0.36. Tests are also conducted by wrapping cigarettes into a semi-circular shape. The bearing capacity and settlement of the footing corresponding to different initial void ratios of the wrapped materials are recorded in Figure 2.12. It can be seen that, under the ultimate load, the settlement increases with the increase in the initial void ratio of the wrapped materials. Also, in comparison to the situation without any reinforcement, the ultimate load can be increased two-fold even by wrapping paper cigarettes to reinforce the foundation. It is worth mentioning that, as well as the extension of the bag, the soilbag becomes flat and tensile force can take place along the bag, so that the wrapped material does not dilate.

Since the ultimate load (not the bearing capacity) is proportional to twice the power of the footing width, it is thus more effective to increase the width $B'$ of the wrapped materials. Therefore, we increase $B'$ to 3B (footing width $B = 10\,\text{cm}$) and 5B (footing width $B = 5\,\text{cm}$). The test results are shown in Figures 2.13 and 2.14, respectively. The theoretical formula shows that the ultimate load would increase about $3^2 = 9$ and $5^2 = 25$ times respectively, in comparison to those

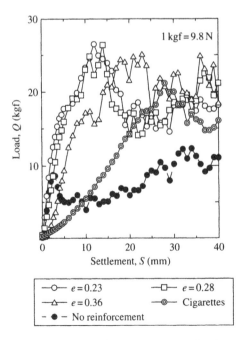

*Figure 2.12* Load Q vs. settlement S relationships in various initial void ratios of the material inside the wrapped part ($B = 10\,\text{cm}$, $B' = 15\,\text{cm}$).

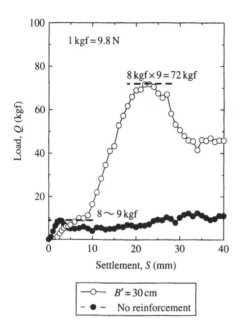

Figure 2.13 Increase in bearing load with the increasing width B′ of the wrapped part (B = 10 cm, B′ = 30 cm).

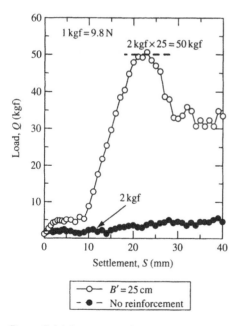

Figure 2.14 Increase in bearing load with the increasing width B′ of the wrapped part (B = 5 cm, B′ = 25 cm).

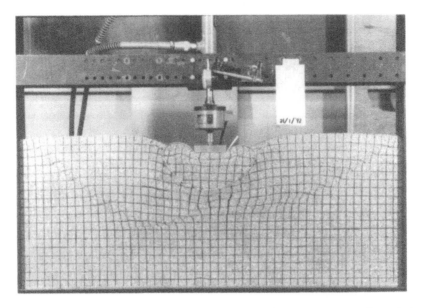

*Figure 2.15* Upward expansion of the wrapped part in two sides of the footing ($B = 10$ cm, $B' = 30$ cm).

without reinforcement. However, we find that, before the ultimate load is reached, the settlement of the footing increases accompanied with the increment of the width of the wrapped material. As illustrated in Figure 2.15, the wrapped material bends upwards and heaves on the two sides of the footing, resulting in significant settlement before the wrapped material is solidified. It is known that soil settlement has detrimental effects on a building, but how do we handle this problem?

We reshape the wrapped material into a 15 cm × 15 cm square (Case 1), in which six layers are divided equally at 15 cm × 2.5 cm (Case 2). A load is applied to the footing. The load ($Q$) and relevant settlement of the footing ($S$) are recorded in Figure 2.16, which corresponds to the slip lines in Figure 2.17 (Case 1) and Figure 2.18 (Case 2), respectively. It may be seen that the slip initially takes place inside the big square (15 cm × 15 cm). When the wrapped material inside the big square has solidified and behaves as part of the footing, another slip occurs beneath the square. Figure 2.16 shows that the $Q - S$ curve of Case 2 has a steeper slope than that of Case 1. This is because less dilation occurs inside each small wrapped body, which in turn leads to less settlement. A larger value of $dQ/dS$ is thus obtained. This suggests that the ultimate load can be increased by subdividing the wrapped material into smaller pieces. The above tests shed light on the re-study of soilbags, although soilbags have been used since ancient times.

We wrap the same amount of aluminum rods in a piece of paper with an initial size of 4 cm × 1.5 cm and arrange them accordingly, as shown in Figure 2.19. These aluminum rod groups are used to simulate the behavior of soilbags in a

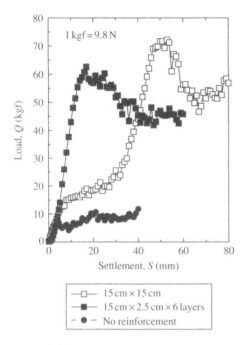

*Figure 2.16* Variation in load Q vs. settlement S relationship after dividing a large wrapped part of 15 cm by 15 cm into 6 smaller ones (B = 10 cm, B' = 15 cm).

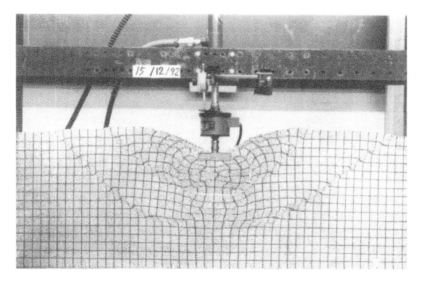

*Figure 2.17* Double slip lines in the model ground with 15 cm square area wrapped material (B = 10 cm, B' = 15 cm).

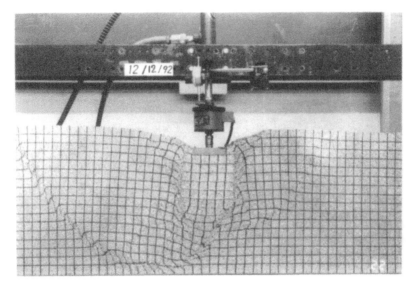

*Figure 2.18* Big slip lines in the model ground by dividing 15 cm square wrapped area into six layers of 15 cm × 2.5 cm (B = 10 cm, B′ = 15 cm).

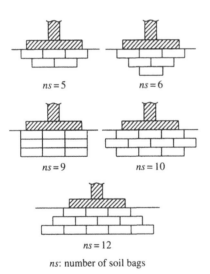

*Figure 2.19* Arrangement of small wrapped parts (model soilbags) (B = 10 cm, each soilbag is 4 cm × 1.5 cm).

two-dimensional manner. A load is applied to the footing. The test results in Figure 2.20 show that the ultimate load may be improved by increasing the number of soilbags. Moreover, the ultimate load is at least double its magnitude in comparison to that without reinforcement by soilbags. This implies that soilbags are effective in foundation reinforcement, regardless of the order of the arrangements.

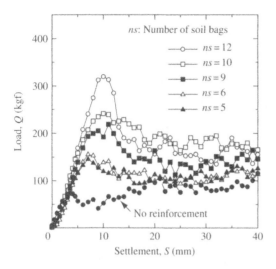

Figure 2.20 Increase of the bearing capacity with the increasing numbers of model soilbags corresponding to Figure 2.19.

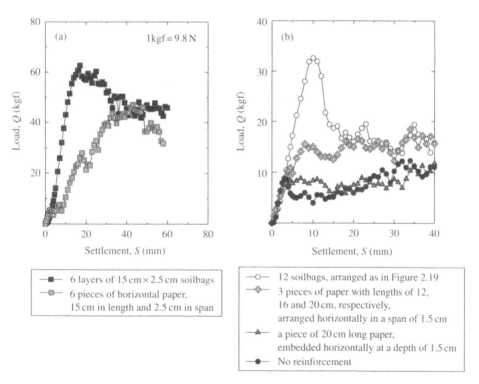

Figure 2.21 Comparison of the reinforcement by soilbags and by horizontally arranged reinforcers (paper).

The commonly employed methods to improve the bearing capacity of foundations mostly place the reinforced materials (such as sheets and grids) horizontally
in the foundation (e.g. Binquest and Lee, 1975a,b; Huang and Tatsuoka, 1990). In
order to make a comparison of the reinforced effects with soilbags, we perform
the following tests by placing layers of papers horizontally in the aluminum rods.
Figure 2.21(a) shows the experimental results in the two situations: the simulated
foundation reinforced by six layers of soilbags (Case 1) and by 6 horizontal layers
of paper (Case 2). It is observed that the ultimate bearing capacity of Case 1 is
higher than Case 2, whilst the corresponding settlement of Case 1 is less than
Case 2. This is because the soilbags are quickly integrated with the footing due
to the dilation of the wrapped particles subjected to external forces. A wider and
deeper foundation is thus formed. In Case 2, some particles escape from two lateral
sides of the reinforced area, resulting in a decreasing width of the reinforced area,
larger settlement and less ultimate bearing capacity. Furthermore, we design the
following three modes of reinforcement. In Case A, 12 soilbags are arranged as
indicated in Figure 2.19. In Case B, 3 pieces of paper with the length of 12, 16
and 20 cm respectively are placed horizontally with a span of 1.5 cm. In Case C,
one piece of paper with the length of 20 cm is horizontally embedded beneath the
foundation surface of 1.5 cm. The test results are shown in Figure 2.21(b), from
which we may see that Case A has the largest value of ultimate bearing capacity.
The above tests show that earth reinforcement by soilbags is more effective and
reliable than some other commonly used methods.

# Chapter 3

# Characteristics of soilbags

## 3.1 Compressive strength and anisotropy

### 3.1.1 Compressive strength in the case of $\delta = 0$

We define the angle between the direction of the major principal stress $\sigma_1$ and the short axis of a soilbag as $\delta$. When soilbags are used to reinforce soft foundation, the soilbags are in general subjected to vertical forces from the upper structures. In this case, $\delta = 0$. Figure 3.1(a) shows a soilbag subjected to the principal stresses $\sigma_{1f}$ and $\sigma_{3f}$ in a two-dimensional manner. The material inside the soilbag is assumed to be frictional and granular. At constant volume condition, under the actions of $\sigma_{1f}$ and $\sigma_{3f}$, the total perimeter of the bag usually increases because the soilbag becomes flatter, as shown in Figure 3.2. Subsequently, the bag becomes flat and a tensile force $T$ is developed along the bag. The dilatancy occurring inside the bag may help to develop high tensile forces. The tensile force $T$ produces an additional stress on the particles inside the bag with components $\sigma_{01} = 2T/B$ and $\sigma_{03} = 2T/H$, in which $B$ and $H$ are the width and height of the soilbag, respectively. As illustrated in Figure 3.1(b), the stresses acting on the particles are thus the combination of the external stresses and the stress caused by the tensile force $T$. At failure, the major principal stress $\sigma_{1f}$ can be calculated by:

$$\sigma_{1f} + \frac{2T}{B} = K_p \left( \sigma_{3f} + \frac{2T}{H} \right)$$

Therefore

$$\sigma_{1f} = \sigma_{3f} K_p + \frac{2T}{B} \left( \frac{B}{H} K_p - 1 \right) \tag{3.1}$$

in which $K_p = (1 + \sin\phi)/(1 - \sin\phi)$ is the lateral earth pressure ratio at passive state. For cohesive-frictional materials, the following relationship exists between the major and minor principal stresses at failure:

$$\sigma_{1f} = \sigma_{3f} K_p + 2c\sqrt{K_p} \tag{3.2}$$

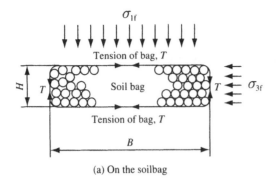

(a) On the soilbag

Tensile force $T$ along the bag is converted to stresses acting on the particles inside the soilbag

$\sigma_{01} = 2T/B$

$\sigma_{03} = 2T/H$

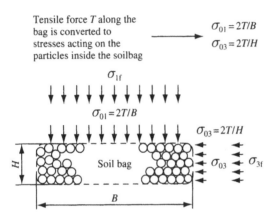

(b) On the particles inside the soilbag: the tensile force along the bag is converted to the stresses acting on the particles inside the soilbag

*Figure 3.1* Force analysis on soilbag and the particles inside the soilbag.

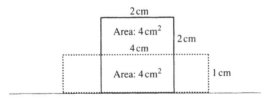

Under the same area of $4\,cm^2$, the perimeter is extended from 8 cm to 10 cm when a 2 cm × 2 cm square becomes a 4 cm × 1 cm rectangle. As a result, a tensile force is produced along the bag

*Figure 3.2* Tensile force is produced along the bag when a soilbag becomes flat.

The apparent cohesion $c$ is thereby expressed as (Matsuoka *et al.*, 2000a):

$$c = \frac{T}{B\sqrt{K_p}}\left(\frac{B}{H}K_p - 1\right) \tag{3.3}$$

Eq. (3.1) is verified through a series of biaxial compression tests on three packages of wrapped aluminum rods. As shown in Figures 3.3 and 3.4, the 5 cm long aluminum rods, with two different diameters of 1.6 and 3 mm, are wrapped with a piece of paper. The mixed rods have a weight ratio of 3:2 and an internal friction angle $\phi = 25°$. Because of the limitation of the loading capacity of our biaxial compression test device, the wrapping paper we used is extremely weak, having the tensile strength $T = 0.36 \text{kgf/cm} = 3.53 \text{N/cm}$. The aluminum package has dimensions of 15 cm wide and 3.75 cm high. The test device allows vertical and lateral loads to be applied separately. To minimize the influence of friction on the top and the bottom of the device, three aluminum packages are piled up vertically to form one specimen. The test results are given in Figure 3.5 through the solid Mohr's stress circles. Moreover, the theoretical estimations through Eq. (3.3), the Mohr's stress diagram (dotted circles) of the wrapped materials at failure, and the two straight lines representing the failure envelopes are also shown in Figure 3.5. The test results clearly show that the wrapped packages exhibit the typical

*Figure 3.3* Setup of the biaxial compression test on three packages of wrapped aluminum rod assemblies.

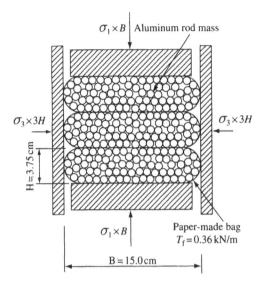

Figure 3.4 Schematic view of the biaxial compression tests.

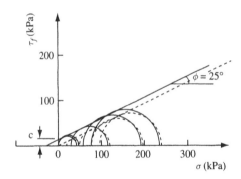

Figure 3.5 Mohr's stress diagram expressing the biaxial compression test results (solid semi-circles). The dotted semi-circles represent the stress diagram of the wrapped particles at failure. The solid straight line is calculated by $\tau_f = c + \sigma \tan 25°$, where $c$ is obtained from Eq. (3.3), and the dotted straight line $\tau_f = \sigma \tan 25°$.

characteristics of cohesive-frictional materials as a whole, whilst the aluminum rods are still frictional material. The phenomenon is beyond common expectations. Generally, a bonding agent (e.g. cement for soil foundations) is used to convert frictional materials to cohesive-frictional materials. Why do the aluminum rods wrapped with a piece of paper show a similar effect if a bonding agent were used? Our understanding is that, due to the constraint of the wrapping paper, a tensile force develops along the paper that leads to an increase in the contact normal force among the wrapped particles. This in turn enhances the friction resistance between particles. Subsequently, the aluminum rods are less likely to slip, displaying behavior as if under the action of a bonding agent. Therefore, the packages of wrapped aluminum rods exhibit very high compressive strength. It is also noted that the predicted failure envelopes (the solid straight lines) in Figure 3.5 are almost tangent to the Mohr's stress (solid) circles at failure, suggesting the reasoning behind Eq. (3.3). This also supports the assumption that, as applied in the derivation of Eq. (3.1), the wrapped material reaches the critical equilibrium state simultaneously with the wrapping paper.

We conduct the unconfined compression tests on soilbags, as shown in Figure 3.6. The bags are usually made of PE, PP and polyester. The PE and PP bags are sensitive to ultraviolet rays, whilst the polyester bags are somewhat resistant to ultraviolet rays and have higher tensile strength. The size of a PE bag is approximately $40\,cm \times 40\,cm \times 10\,cm$ and the polyester bag is $38\,cm \times 30\,cm \times 8.5\,cm$.

*Figure 3.6* Unconfined compression test on soilbags.

The materials inside the bags consist of crushed stones with internal friction angle $\phi = 44°$. Through the unconfined compression tests, we measure the bearing capacity for various types of bags filled with crushed stones. The maximum load of average PE soilbags ranges from 230 to 280 kN (the equivalent bearing capacity being 1400–1800 kPa), whilst the maximum load of the special polyester bags is 540–640 kN (with an equivalent bearing capacity of 4700–5600 kPa). It is surprising to see the maximum load in the PE soilbags even up to 230–280 kN. The typical loading settlement curve measured in the unconfined compression tests is illustrated in Figure 3.7. Considering $\sigma_3 = 0$ for unconfined compression tests, the maximum load $F$ of a soilbag can be calculated by using Eq. (3.1) as:

$$F = \sigma_{1f} \times B \times L = (2T/B)\left\{(B/H)\,K_p - 1\right\} \times B \times L = 200\,\text{kN} \tag{3.4}$$

in which $L$ is the length of the soilbag. The calculated value of 200 kN is a little smaller than the measurements 230–280 kN, probably caused by the following major factors:

a   Eq. (3.1) only considers the major and minor principal stresses $\sigma_1$ and $\sigma_3$, neglecting $\sigma_2$;
b   Initial dimensions of the soilbags are used in the calculation through Eq. (3.4) instead of the actual size at failure.

Nevertheless, the calculated maximum load of 200 kN is 40 times higher than the tensile strength of the bag itself (4.7 kN), which is greatly beyond normal expectations. Analogously, the high bearing capacity of the soilbags is due to the apparent cohesion, which can be traced from the tensile force developed along the bag. It is interesting to see that soilbags have the ability to convert the external force which is the "enemy" of the foundations into the "friend" of the foundations

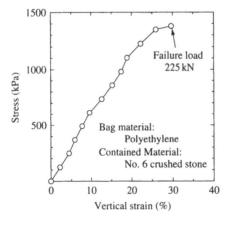

*Figure 3.7* Typical stress–strain relationship of soilbags obtained from the unconfined compression tests.

due to the action of the tensile force developing along the bags. This is the "reversal idea" involved in the mechanism of soilbags. The above tests also demonstrate that the bearing capacity of soilbags can be increased considerably if high tensile strength materials like polyester are adopted for the bags.

### 3.1.2 Strength anisotropy in the case of $\delta \neq 0$

Soilbags exhibit high compressive strength when subjected to external forces (the major principal stress) along the short axis of the soilbags. In practice, however, soilbags may undertake external forces from any direction. The compressive strength may decrease with the increase in inclination between the external forces and the short axis of the soilbag. We investigate this strength anisotropy through a series of biaxial compression tests. As shown in Figure 3.8, the test apparatus allows vertical and horizontal loads to be applied on the samples independently. Two different kinds of aluminum rods (1.6 and 3 mm in diameters, 50 mm long and the weight ratio of the mixed assembly 3:2) are wrapped with a piece of weak paper (tensile strength 8.24 N/cm) to construct a package of 5 cm wide by 1 cm high. These packages are piled up with an inclination ($\delta$) of 0°, 15°, 30°, 45°, 60° and 90° to the horizontal direction. Figure 3.9 shows the measured Mohr's stress diagrams at failure in the conditions of $\delta = 0°$, 15°, 30° and 45°. The apparent cohesions corresponding to different inclinations, denoted as $c(\delta)$, can be obtained from the tangent lines of these Mohr's stress diagrams. The values of $c(\delta)$ are normalized by the value of $c$ at $\delta = 0$, as illustrated in Figure 3.10. We also

*Figure 3.8* Biaxial compression tests on the wrapped aluminum rod assemblies.

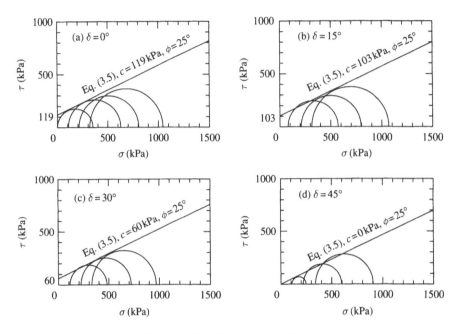

Figure 3.9 Mohr's stress diagram showing the biaxial compression test results under different inclinations of the piled packages (the wrapped aluminum rod assemblies). The solid straight line is calculated by $\tau_f = c + \sigma \tan 25°$ where c is obtained from Eq. (3.3).

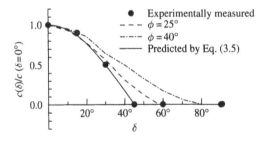

Figure 3.10 Variations of the apparent cohesions with the inclination of the wrapped aluminum rod assemblies.

calculate the $c$ value through the formula by Chen (1999) when $\phi = 25°$ and $40°$, denoted by the dotted lines in Figure 3.10. The dotted lines show that $c(\delta)$ increases with the increasing internal friction angle $\phi$ of the wrapped materials. Moreover, the measurements can be fitted with the following expression:

$$c(\delta) = \begin{cases} c(\delta = 0) \cdot \cos 2\delta & (0° \le \delta \le 45°) \\ 0 & (45° \le \delta \le 90°) \end{cases} \tag{3.5}$$

where $c(\delta = 0)$ is calculated as:

$$c(\delta = 0) = \frac{T}{B\sqrt{K_p}}\left(\frac{B}{H}K_p - 1\right)$$

(3.6)

in which $B$ and $H$ are the width and the height of the wrapped aluminum rods, $K_p$ is the ratio of passive earth pressure and $T$ is the tensile strength of the wrapping paper. It is obvious that Eq. (3.6) is identical to Eq. (3.3). It is also seen from Figure 3.10 that the measurements are on the safe side in comparison to the calculations. Moreover, taking the internal friction angle of the aluminum rods as $\phi = 25°$, the cohesions $c(\delta)$ are calculated through Eq. (3.5) when $\delta = 0°$, 15°, 30° and 45°. The failure envelopes are thus obtained (shown in Figures 3.9(a–d), respectively). It is seen that the apparent cohesion $c(\delta)$ decreases with the increase in $\delta$, and it becomes zero when $\delta \geq 45°$.

## 3.2  Vibration reduction

### 3.2.1  Laboratory cyclic simple shear tests

In the cyclic simple shear tests apparatus (Figure 3.11), the vertical load is applied through an oil cylinder, and the horizontal (shear) load is created by pulling or pushing the top rigid platen by another oil cylinder. A mobile rhomboid-shaped steel frame connects the top rigid platen to the upper beam of the apparatus, enabling the top rigid platen to move horizontally without rotation. The tested soilbags are filled with two kinds of sands: No. 6 and No. 3 silica sands with an average grain size of 0.25 and 1.2 mm, respectively (Yamamoto et al., 2003). The tests are performed by cyclically moving the top rigid platen under a constant strain.

Figure 3.11 Cyclic simple shear tests on soilbags.

The vertical stresses, taken as 130 and 310 kPa respectively, are kept constant during the tests. The tested soilbags are piled up vertically, Nos 3 and 6 silica sands as two groups (Figure 3.12). The stress–strain relationships of the assemblies are recorded in Figures 3.13–3.15. For comparison, cyclic simple shear tests are also carried out on the Nos 3 and 6 silica sands (Figure 3.16). The sand specimen, filled in a cell (7 cm in diameter, 2 cm in height), has a void ratio of 0.94. Under $\sigma_3 = 130$ kPa, the typical stress–strain relationships of the sands are shown in Figure 3.17.

In order to evaluate the sand behavior under cyclic shear, we define an equivalent damping ratio $h_{eq}$ as:

$$h_{eq} = \frac{1}{2\pi} \frac{\Delta W}{W} \qquad (3.7)$$

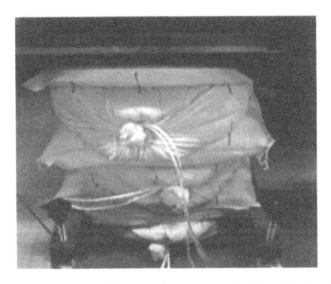

*Figure 3.12* Six soilbags piled up vertically in the cyclic simple shear tests.

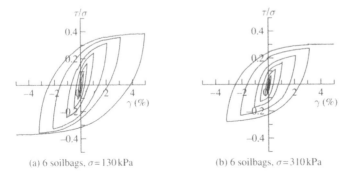

(a) 6 soilbags, $\sigma = 130$ kPa  (b) 6 soilbags, $\sigma = 310$ kPa

*Figure 3.13* Under different vertical stresses, the cyclic stress–strain relationships for the assembly of six soilbags filled with No. 6 silica sand.

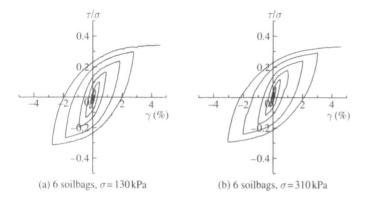

(a) 6 soilbags, $\sigma = 130$ kPa          (b) 6 soilbags, $\sigma = 310$ kPa

*Figure 3.14* Under different vertical stresses, the cyclic stress–strain relationships for the assembly of six soilbags filled with No. 3 silica sand.

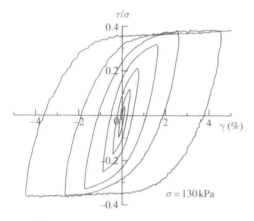

The wrapped materials: No. 3 silica sand

*Figure 3.15* Under $\sigma = 130$ kPa, the cyclic stress–strain relationships for the assembly of three soilbags filled with No. 3 silica sand.

where $W$ is the elastic strain energy and $\Delta W$ is the loss of energy in one cyclic load. The larger area of the stress–strain loop corresponds to a higher equivalent damping ratio $h_{eq}$. The equivalent damping ratios $h_{eq}$ of the soilbags and the sand samples at different shear strains $\gamma$ are calculated by Eq. (3.7) and shown in Figure 3.18. It is seen that, when the shear strain $\gamma$ is higher than 1%, the values of $h_{eq}$ of the soilbags are slightly higher than those of the sand samples. When $\gamma$ is less than 1%, the values of $h_{eq}$ of the soilbags are much higher than those of the sand samples. In general, soils have high equivalent damping ratios at large shear strains. When soils are reinforced, like the concrete-reinforced foundations, the stiffness of the soils increases and the shear strain becomes small under loads. This leads to the decrease in $h_{eq}$. During earthquakes, the shear strain of the soil

*Figure 3.16* Cyclic simple shear tests on Nos 3 and 6 silica sands.

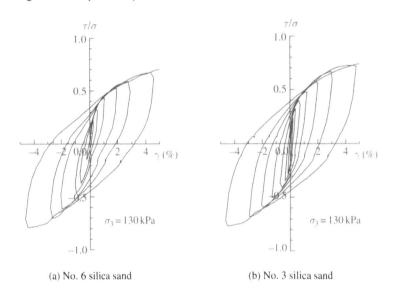

(a) No. 6 silica sand                    (b) No. 3 silica sand

*Figure 3.17* Cyclic stress–strain relationships for the sand samples.

is usually small. In this case, the values of the damping ratio are within a narrow band, as seen in Figure 3.18(b). This is positive for the stability of buildings. However, the above tests show that the assembly of soilbags has higher $h_{eq}$ even under small shear strain. Furthermore, Figure 3.18(b) demonstrates that the assemblies of Nos 3 or 6 soilbags have similar $h_{eq}$, suggesting the independence of the $h_{eq}$ on the numbers of soilbags. Also, the equivalent damping ratio of the assembly of soilbags $h_{eq} = 0.3$ is much higher than those of both the concrete structures

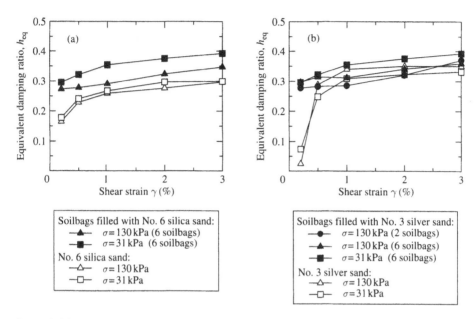

Soilbags filled with No. 6 silica sand:
  —▲—   $\sigma = 130\,kPa$ (6 soilbags)
  —■—   $\sigma = 31\,kPa$  (6 soilbags)
No. 6 silica sand:
  —△—   $\sigma = 130\,kPa$
  —□—   $\sigma = 31\,kPa$

Soilbags filled with No. 3 silver sand:
  —●—   $\sigma = 130\,kPa$ (2 soilbags)
  —▲—   $\sigma = 130\,kPa$ (6 soilbags)
  —■—   $\sigma = 31\,kPa$  (6 soilbags)
No. 3 silver sand:
  —△—   $\sigma = 130\,kPa$
  —□—   $\sigma = 31\,kPa$

*Figure 3.18* Equivalent damping ratios of the soilbags and the sand samples.

($h_{eq} = 0.05$) and the steel structures ($h_{eq} = 0.02$). The $h_{eq}$ value of soilbags is close to that of rubber ($h_{eq} = 0.1–0.3$). This is convenient in the foundation reinforcement design, using soilbags as a vibration absorber.

### 3.2.2  Laboratory vibration tests

Vibration tests are conducted on five soilbags piled up vertically, as shown in Figures 3.19(a) and (b). Each soilbag is 40 cm in width and 8 cm in height. The soilbags are filled with No. 3 silica sand with maximum, average and minimum grain sizes of 3.4, 1.2 and 0.3 mm, respectively. Four acceleration sensors of small strain gauges are respectively set on the interfaces between soilbags. The lowest sensor only receives 8.8% of the total input acceleration, and 13% of the value of the highest sensor. If the same silica sand is used to fill a paper-made box with the same height as the three soilbags piled up vertically (Figure 3.19(c)), the lowest sensor receives 12.6% of the total input acceleration, and 57% of the value of the highest sensor. Moreover, if the same silica sand is cumulated without any lateral restrictions (Figure 3.19(d)), the lowest sensor receives 14.6% of the total input acceleration, and 39% of the value of the highest sensor. It is obvious that the piled soilbags are the most effective way to reduce the input acceleration. Under loading and unloading conditions, the soilbags have a little compressive and rebounding deformation, respectively. Simultaneously a slight extension and contraction occurs on the bags. The vibration energy is thus absorbed in the frictional materials such as by the sand in the bags.

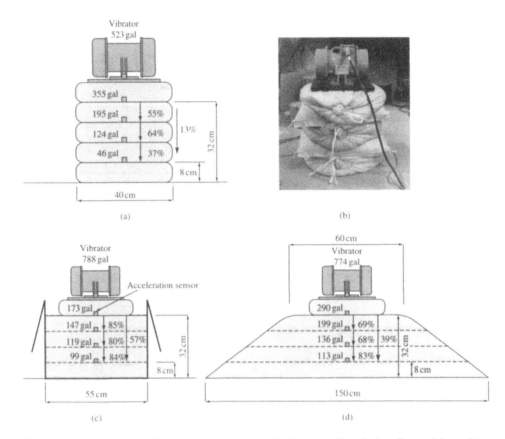

*Figure 3.19* Laboratory vibration tests on (a) and (b) vertically piled soilbags, (c) sand in a paper-made box, and (d) sandbank to investigate the effect of downward vibration.

Furthermore, we investigate the effects of vibration reduction along the vertical direction through the adjacent soilbags (Matsuoka *et al.*, 2003b, 2004b). The test design is shown in Figure 3.20, where two columns of vertically piled soilbags contact laterally (Cases (a) and (c)) or are separate (Cases (b) and (d)). We install two types of platen vibrators with vibration frequencies of 60 Hz (Cases (a) and (b)) and 6 Hz (Cases (c) and (d)) on one column of soilbags. In Cases (b) and (d), two columns of the piled soilbags are arranged separately, the accelerations on the adjacent columns of soilbags are recorded and shown in Figures 3.20(b) and (d). Considering the tolerance of the strain-typed gauges is around ±10 gal, the accelerations transmitted to the adjacent column of soilbags in Cases (a) and (c) are almost zero, which suggests the vibration is difficult to pass to the adjacent columns, although the two columns contact laterally. This is because the vibration transmits to the neighboring columns through the floor rather than through lateral contact of soilbags. Therefore, the lateral contact of the two columns of soilbags does not play an important role in vibration transmission. The phenomenon can also be observed through the fluctuation of

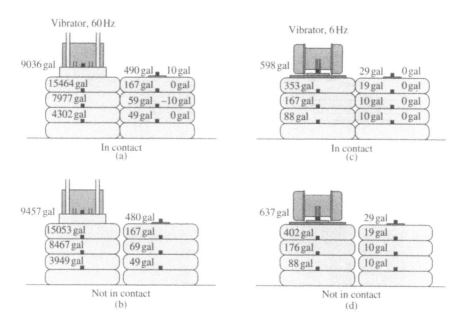

*Figure 3.20* Laboratory tests on real soilbags to investigate the transmission of vibration to the adjacent column of soilbags.

two (glass) cups of water separately sitting on the top of two columns of soilbags. In Case (a) where a cup of water is placed on the column with a vibrator, the water level fluctuates significantly when the vibrator runs. However, the water level is almost at rest when the cup is placed on the adjacent column without the vibrator. Besides, the vibration can also be felt by touching the soilbags by hand.

### 3.2.3  In situ *vibration tests*

The schematic view of the *in situ* vibration test is shown in Figure 3.21, where a pit of 280 cm wide, 280 cm long and 56 cm deep is filled with soil bags (Matsuoka *et al.*, 2004b). Each soilbag is 40 cm wide, 40 cm long and 8 cm high. Four layers of soilbags are piled up vertically up to a total height of 32 cm as the foundation. As a barrier against vibration, another 3 layers of soilbags are constructed above the foundation on the right side of the pit, piling up vertically to a height of 24 cm. The widths of the bottom and top layers of the barrier are 160 and 80 cm, respectively. The left side of the pit is backfilled with the same soils as used in the soilbags with a grain size of less than 25 mm. At the center of the left side of the pit (the portion of sand backfilled), two types of vibrators are installed to produce vibration: one is the electrically driven platen vibrator with a frequency of 60 Hz and a weight of 25 kg; another is the engine-driven platen vibrator with a frequency of 90 Hz and a weight of 60 kg. Data are collected at every 40 cm along the line perpendicular to the soilbag barrier. The measured vibration levels in decibels are

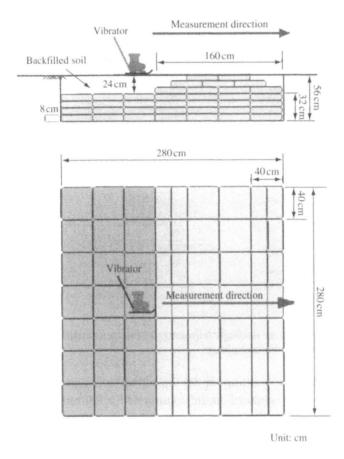

Unit: cm

*Figure 3.21* Schematic plan view and typical cross-section of the *in situ* test in a pit constructed with soilbags. The left side of the pit is backfilled with soils above four layers of soilbags. Vibrators are installed in the center of the surface of the backfilled soils.

plotted respectively in Figure 3.22 for the electrically driven platen vibrator and in Figure 3.23 for the engine-driven platen vibrator, in which the solid lines are the distance damping against vibration measured near the pit. It can be seen that the vibration is reduced significantly through the soilbag barrier in comparison to the distance damping. The average vibration reduction is about 15 dB for the electrically driven platen vibrator and 10 dB for the engine-driven platen vibrator.

### 3.2.4  Case history study

#### 3.2.4.1  Applications in truck roads

Soilbags have already been applied to reinforce truck roads in Nagoya city with asphalt and concrete pavements. The following are two representative cases.

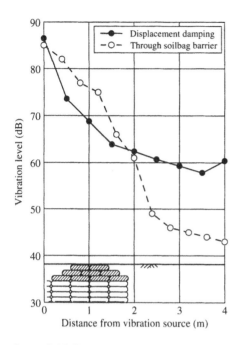

*Figure 3.22* The measured vibration levels in decibels (dB) vs. the distances from the vibration source of the electrically driven platen vibrator.

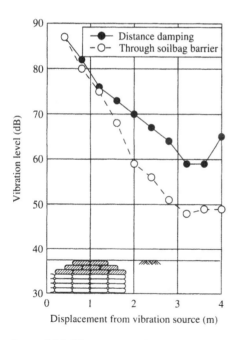

*Figure 3.23* The measured vibration levels in decibels (dB) vs. the distances from the vibration source the engine-driven platen vibrator.

*Figure 3.24* Vibration reduction work using soilbags to reinforce the truck road foundation in Nagoya City.

The first case is a 20 m wide road, where the asphalt pavement has been severely damaged due to the weak roadbed beneath and the heavy traffic. Additionally, the nearby residents have been suffering from the traffic-induced vibration. In the reconstruction of this road, as shown in Figure 3.24, the soilbags are placed under the truck road with a cover of asphalt pavement. The cross-section is illustrated in Figure 3.25. Three layers of soilbags with a total height of about 20 cm are constructed as the foundation, above which is a 15 cm thick layer of recycled crushed stone (RC-40). The crushed stone is used to protect the underlying soilbags from the high temperature of the asphalt concrete during paving. In order

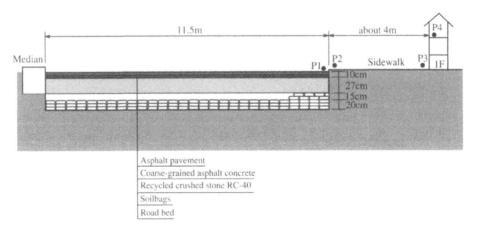

*Figure 3.25* Cross-section of the truck road foundation reinforced with asphalt pavement and soilbags, in which P1 to P4 are the measurement points of vibration.

*Table 3.1* The vibration levels at the asphalt-paved truck road measured before and after the reinforcement by soilbags

| Locations of measurements | P1 | P2 | P3 | P4 |
|---|---|---|---|---|
| Before reinforcement | 67 dB | 66 dB | 57 dB | 65 dB |
| After reinforcement | 55 dB | 54 dB | 46 dB | 50 dB |

to effectively reduce the traffic-induced vibration, 2 layers of soilbags are placed near the sidewalk. Furthermore, a 27 cm thick layer of coarse-grained asphalt concrete is constructed, followed by 10 cm thick asphalt as pavement. Before and after the reinforcement with soilbags, we measure the vibration at the sidewalk (points P1 to P3) and at the third floor of a steel-framed house (point P4). The measurements are listed in Table 3.1, in which the data are obtained by averaging the maximum vibration levels within 10 minutes during the measurements. It can be seen that, after the reinforcement by soilbags, the vibration levels at the four locations (points P1 to P4) are all less than 60 dB, which is the highest permissible vibration level for road traffic vibration at night. In particular, the residents in the steel-framed houses were satisfied after the reconstruction of the road because vibrations were brought down to an acceptably low level.

Another case of truck road reinforcement using soilbags but with under-concrete pavement is illustrated in Figures 3.26 and 3.27. Table 3.2 summarizes the average maximum vibration levels measured at two ends of the sidewalk (points P1 and P2) before and after the reconstructions. Similarly, the vibration levels are reduced considerably after the reinforcement. It may be summarized that, either under asphalt or concrete pavements, the traffic-induced vibration can be significantly reduced after reinforcement with soilbags. Furthermore, vibration reduction with soilbags will be verified through long-term vibration measurement at the above two sites.

### 3.2.4.2   Applications in soft building foundations

When soilbags are applied to the reinforcement of soft foundations, soilbags not only improve the bearing capacity of foundations, but also reduce traffic-induced vibrations. The following are two examples on the subject (Matsuoka *et al.*, 2003b, 2004b; Matsuoka and Liu, 2003).

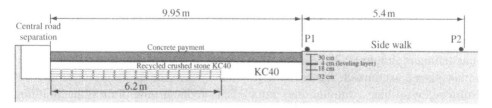

*Figure 3.26* Cross-section of the reinforced truck road with concrete pavement and soilbags, in which P1 and P2 are the locations of vibration measurements.

*Figure 3.27* Reinforcement of the truck road with concrete pavement and soilbags.

*Table 3.2* Vibration levels of the concrete-paved truck road measured before and after reinforcement with soilbags

| Locations of measurement | P1 | P2 |
|---|---|---|
| Before reinforcement | 69 dB | 66 dB |
| After reinforcement | 57 dB | 54 dB |

*(1) Vibration measurement in YM city, Kanagawa Prefecture, Japan*
The foundations to be reinforced with soilbags in YM city, Japan, mainly consist of black clay (volcanic cohesive soil). The soil is so weak that people cannot walk on it. Because the black clay in the field is difficult to dispose of, it is put into the bags to form soilbags. These soilbags are arranged in the foundation under the building (Figure 3.28) and compacted layer by layer using small vibrators and wood hammers (Figure 3.29). As shown in Figure 3.30, one vibration measurement point is placed just outside the building (point P1) where the underground soil is not reinforced by soilbags. Another one (point P2) is on the ground floor of the building where the foundation has been reinforced with soilbags. The directions along and perpendicular to the traffic road are defined as $x$- and $y$-directions, respectively. The

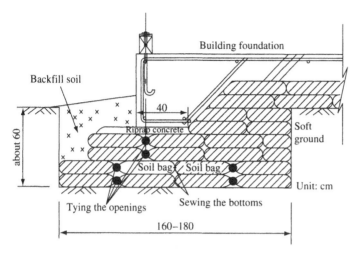

Figure 3.28 Schematic view of the cross-section of the soft foundation $(N = 1-2)$ reinforced with soilbags.

Figure 3.29 Soft foundation reinforcement with soilbags in YM city, Kanagawa Prefecture, Japan.

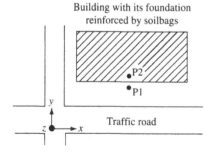

Figure 3.30 Locations of the vibration measurement points (P1 and P2) along x-, y- and z-directions.

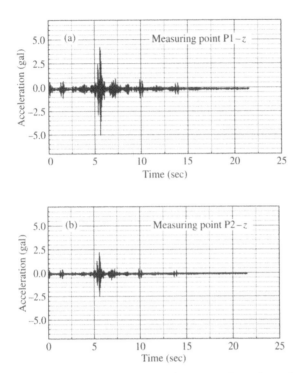

*Figure 3.31* The measured accelerations along the z-direction (a) P1: outside the building (no reinforcement on the foundation), and (b) P2: inside the building
(foundation reinforced).

upward vertical direction is the $z$-direction. The vibration acceleration distributions
along the $z$-direction at points P1 and P2 are shown in Figure 3.31. It can be
seen that the traffic-induced vibration accelerations decrease significantly after
passing through the reinforced foundation. In order to investigate the frequency
distributions where the vibration accelerations reduce the most, the acceleration
spectrums at points P1 and P2 are analyzed within every vibration frequency of
0.2 Hz. The ratios of the acceleration spectrum in $z$-, $x$- and $y$-directions vs the
vibration frequency are recorded in Figure 3.32. It may be seen that, within the
vibration frequency of 1–10 Hz, the acceleration spectrums at point P2 are less
than half of the values at point P1 in the three directions (the ratio of acceleration
spectrum is about 0.4). It is understood that the frequency of vibration between 5 to
8 Hz is the range most sensitive to human feeling. Besides, the resonance frequency
of a wood-made building is usually less than 10 Hz. Therefore, soilbags are suitable
to apply to the foundation reinforcements of low- or medium-height buildings.

*(2) Vibration measurement in YC cho, Ibaraki Prefecture, Japan*
The foundation soil in YC cho, Japan, is very weak, similar to that in YM city,
which we described in the previous case. Considering the similar profiles of the

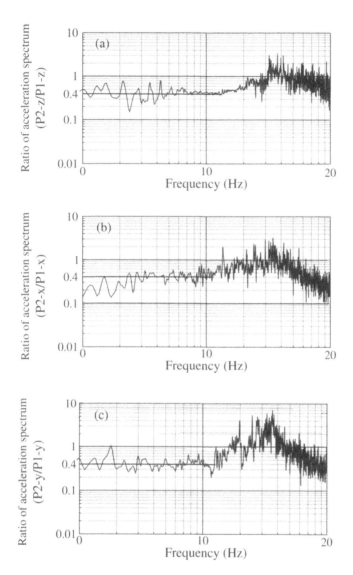

*Figure 3.32* Relationships between the ratio of the acceleration spectrum and the frequencies in the x-, y- and z-directions, where P1 is outside the building (no reinforcement on the foundation), and P2 is inside the building (foundation reinforced).

two foundations, we conduct a similar design to reinforce the present foundation, as illustrated in Figure 3.28. The construction processes of the soilbags and the raft foundation are recorded in Figures 3.33 and 3.34. The locations for the vibration measuring points P1 and P2 are similarly installed as those in Figure 3.30, except that the traffic road is about 20–30 m away from the buildings. The directions along and perpendicular to the traffic road are defined as *x*- and *y*-directions,

*Figure 3.33* Soft foundation reinforcement under the footing of soilbags in YC cho, Ibaraki Prefecture, Japan.

*Figure 3.34* Two layers of soilbags are placed beyond the footing. A steel-reinforced raft foundation is constructed above the soilbags.

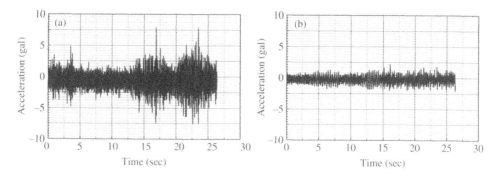

*Figure 3.35* Accelerations along the z-direction: (a) outside the building (no reinforcement on the foundation), and (b) inside the building (with reinforcement on the foundation).

respectively. The upward vertical direction is the $z$-direction. The vibration accelerations of points P1 and P2 in the $z$-direction are plotted in Figure 3.35. In comparison with Figures 3.35(a) and (b), the acceleration of the traffic-induced vibration decreases significantly when the vibration passes through the reinforced foundation. As the main objective of using soilbags is to improve the bearing capacity of a soft foundation, the reduction of the traffic-induced vibration is an additional benefit. The mechanism of the reduction of vibration by soilbags has not yet been completely clarified, although we believe that one contribution may be due to the different degrees of stiffness of the solid soilbags and the surrounding soft foundation, which causes the loss of vibration input. Moreover, the discontinuity of soilbags may also absorb vibration energy or reduce vibration.

Here we would like to introduce an interesting story on vibration reduction using soilbags. In Chapter 5 (Figure 5.20), we will describe the details of the foundation design of a construction site near the JR (Japan Railway) line at the TK city of Hokaido, Japan. We briefly introduce an interesting test on the reduction of vibration using soilbags. At this construction site, the foundation soil has a low $N$-value of between 2 and 3, and the foundation thereby needs to be reinforced. In our design plan, a steel-reinforced raft foundation is constructed above the soilbags. When the foundation reinforcement is almost complete, three cups of water are separately placed: cup I on the soilbags, cup II on the short concrete and cup III on the ground, as illustrated in Figure 3.36. These three cups are an equal distance from the JR line. When a train passes, the water level of cup III fluctuates, whilst the water levels of cups I and II do not fluctuate at all. This simple test provides an intuitive understanding that reinforcement with soilbags also benefits the reduction of traffic-induced vibration.

*Figure 3.36* Verification of the reduction of traffic-induced vibration through the observation of the water level fluctuations in three cups. At the same distance from the JR line, three cups of water are placed on (a) the top of the soilbags, (b) the surface of the reinforced foundation, and (c) the surface of the foundation without reinforcement.

## 3.3 Frost heave prevention

Frost heave is a phenomenon where the ground soils expand in cold regions due to the freezing of water. If there is a continuous supply of ground water, mainly due to capillary action, frost heave easily occurs. Frost heave causes the settlement of foundations, which can thereafter lead to cracking on the upper buildings. The conventional method of preventing frost heave is to deepen the concrete footings over the specified depth of frost heave (Figure 3.37), which usually results in a long construction period and vast construction costs. We thereby propose a new method to prevent frost heave using soilbags filled with coarse materials (Figure 3.38). The soilbags are placed within the depth in which frost heave would usually occur. The number of layers of soilbags, usually from 4 to 10, is determined through the specified depth for frost heave prevention and the bearing capacity of the ground. Since the soilbags filled with coarse materials have large voids, the capillary water finds it difficult to rise inside the soilbags as well as through the soil ground above the soilbags. Suzuki *et al.* (2000) carried out frost heave prevention tests by embedding soilbags into the ground at a depth of 20 cm, where the ground was volcanic cohesive clay. The soilbags were filled with

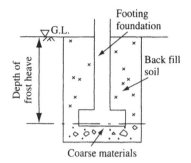

*Figure 3.37* Conventional method of frost heave prevention.

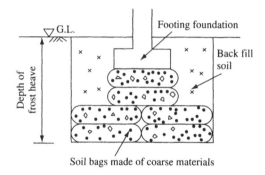

*Figure 3.38* The proposed method of frost heave prevention using soilbags filled with coarse materials.

crushed stone with grain sizes of 2.5–20 mm. They found that the water content of the soils (20 cm above the soilbags) increased only 5%. This suggests that the capillary water hardly rose inside the soilbags and thereby frost heave would not take place. Without soilbags embedded, the increase in water content in the surface of the ground (also 20 cm above the soilbags) was up to 20%. The effectiveness of frost heave prevention using soilbags filled with coarse materials is evident. So far, this new method has been applied in the Hokkaido area (in the northern regions of Japan).

## 3.4 Tensile strength

Commonly, there is no tensile strength for frictional materials. However, if the frictional materials are put into a bag to form a soilbag, the soilbag can withstand tensile forces. In other words, the soilbag has tensile strength. This has been verified through a series of pulling tests as shown in Figure 3.39 (Matsuoka *et al.*, 2002a; Matsuoka and Liu, 2003). We use similar aluminum rods and wrapping paper as those used in the biaxial compression tests (Section 3.1) to construct a soilbag model. Under different width ($B$) and height ($H$), the Mohr's stress

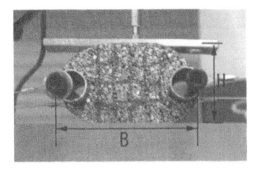

*Figure 3.39* Pulling the wrapped aluminum rod assemblies horizontally without vertical loading.

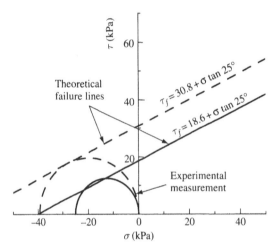

Cohesions used in theoretical failure lines are predicted
by Eq. (3.3), under conditions of
$B = 6.3$ cm, $H = 4.1$ cm, $T = 10.7$ N/cm for $c = 30.8$ kPa,
$B = 9.4$ cm, $H = 5.2$ cm, $T = 8.0$ N/cm for $c = 18.6$ kPa

*Figure 3.40* Mohr's stress diagrams obtained from the pulling tests and the theoretical prediction.

diagrams of the soilbag model at failure (when the wrapping paper is broken) are shown in Figure 3.40. The failure envelopes calculated by Eq. (3.3) are also plotted for comparison. It can be seen that the Mohr's stress diagram varies with the size of the soilbag and the tensile strength of the wrapping paper. Moreover, the failure envelopes calculated by Eq. (3.3) are basically tangent to the Mohr's stress diagrams, suggesting that the tensile strength of the soilbag can be predicted by Eq. (3.3). Nevertheless, the tensile strength of soilbags is usually ignored in practice for the sake of safety, although it may exist when soilbags are connected and used to construct retaining walls.

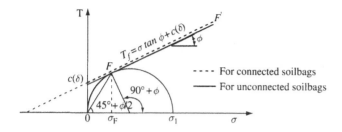

*Figure 3.41* Failure criterion of soilbags.

## 3.5  Failure criterion

Due to the above tests, we may conclude that soilbags have both compressive and tensile strength when they are connected. The compressive strength decreases with the decrease of the apparent cohesion $c(\delta)$ when soilbags are subjected to inclined external forces $(\delta > 0)$. The failure criterion of soilbags in practical design is summarized in Figure 3.41, in which the Mohr's stress diagram represents the stress states of soilbags at failure when $\sigma_3 = 0$, and $c(\delta)$ is calculated by Eqs (3.3) and (3.5). For the connected soilbags, the failure criterion may be considered as $\tau_f = c(\delta) + \sigma \tan \phi$, shown as a bold dotted straight line in Figure 3.41. For the sake of safety, only the compressive strength in the range of $\sigma \geq 0$ is considered in practical applications. Since tensile force cannot be applied externally on unconnected soilbags, the tensile strength of the unconnected soilbags is thus equal to zero in the range of $\sigma_3 < 0$. When $\sigma_1 = \sigma_3 = 0(\sigma = 0)$, no tensile force occurs along the bags, resulting in $\tau_f = c = 0$. Therefore, the strength line should pass the origin of the coordinate for the unconnected soilbags. When $\sigma_3 \geq 0$, the normal stress on the failure plane $\sigma \geq \sigma_F$, where $\sigma_F$ is the normal stress on the failure plane in accordance with $\sigma_3 = 0$. The strength of the unconnected soilbags is characterized by the straight line FF' with an expression of $\tau_f = c(\delta) + \sigma \tan \phi$. In the range of $0 < \sigma \leq \sigma_F$, the strength of the unconnected soilbags increases along OF until the bags reach the limit of their tensile strength. Therefore, the failure criterion of the unconnected soilbags may be characterized by OFF' (the solid line) in practical applications.

## 3.6  Deformation

The deformation characteristic of soilbags is important in design and construction. We propose a method to predict the deformation of soilbags and verify it through biaxial compression and unconfined compression tests (Matsuoka *et al.*, 2003a, 2004a).

### 3.6.1  Deformation estimation when soilbags are subjected to major principal stress along the short axis of soilbags $(\delta = 0)$

The stresses are applied on a soilbag as shown in Figure 3.1 (Section 3.1). $\sigma_1$ and $\sigma_3$ denote the major and minor principal stresses externally applied to the soilbag, respectively. $\sigma_{1m}$ and $\sigma_{3m}$ are the major and minor principal stresses acting on the materials inside the bag. $T$ is the tensile force along the bag developing from zero to its tensile strength. The height and width of the soilbag are symbolized as $H$ and $B$, respectively. The initial height and width of the soilbag are $H_0$ and $B_0$, respectively. The ratio of the initial height to the initial width of the soilbag is defined as $n = B_0/H_0$. In the equilibrium state, the forces in vertical and horizontal directions satisfy the following equations:

$$\left. \begin{array}{l} \sigma_1 + \frac{2T}{B} = \sigma_{1m} \\ \sigma_3 + \frac{2T}{H} = \sigma_{3m} \end{array} \right\} \tag{3.8}$$

For the materials inside the soilbag, the principal stress ratio of $\sigma_{1m}/\sigma_{3m}$ is related to the major principal strain $\varepsilon_1$ by:

$$\frac{\sigma_{1m}}{\sigma_{3m}} = f(\varepsilon_1) \tag{3.9}$$

Eq. (3.9) may be determined through triaxial compression tests on the same materials with the same density as those inside the soilbag. However, difficulties may arise in performing triaxial compression tests due to the large grain size of the tested materials. Alternatively, two other possible ways may be used to determine the expression of Eq. (3.9). The first is numerical methods using elasto-plastic constitutive models of soils like the Cam-clay model. However, it is difficult to handle actual engineering soils due to the conflict between the complexity and variety of soils and the idealization of the constitutive models. Another way is to assume an exponential function to approach Eq. (3.9), which method will be adopted here. We assume the following exponential function:

$$\frac{\sigma_{1m}}{\sigma_{3m}} = a \exp(-100\varepsilon_1) + K_p \tag{3.10}$$

where $K_p = (1 + \sin\phi)/(1 - \sin\phi)$ is the passive pressure coefficient for the materials inside the soilbags, and factor $a$ is determined based on the initial state of the materials inside soilbags, for example $a = 1 - K_p$ for the isotropic consolidation state of $\varepsilon_1 = 0$ and $\sigma_{1m}/\sigma_{3m} = 1$. Figure 3.42 shows the relationship between the principal stress ratio $\sigma_1/\sigma_3$ and the major principal strain $\varepsilon_1$ for the Toyoura sand, in which the solid curve is obtained from the triaxial compression test and the dashed curve is calculated from Eq. (3.10) by considering $\phi = 40°$. The dashed straight line in Figure 3.42 corresponds to the unloading–reloading conditions, taking the initial slope of the dashed curve at $\varepsilon_1 = 0$.

The strain $\varepsilon_y$ along the short axis of the soilbag is calculated by

$$\varepsilon_y = \frac{H_0 - H}{H_0} \tag{3.11}$$

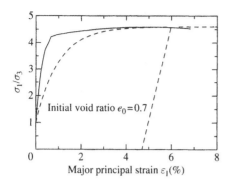

Figure 3.42 Stress–strain relationship of the Toyoura sand-filled soilbags.

If the $\sigma_1$-direction coincides with the short axis of the soilbag ($\delta = 0$), then $\varepsilon_y = \varepsilon_1$, as illustrated in Figure 3.43. Meanwhile, the pulling tests on the bags give the relationship between the tensile force $T$ and the strain $\varepsilon$ of the bag:

$$T = k\varepsilon \tag{3.12}$$

where $k$ is the bag material coefficient with the unit of N/cm. The strain $\varepsilon$ of the bag is obtained from the initial perimeter $L_0 = 2(B_0 + H_0)$ and the instant perimeter $L = 2(B + H)$ of the bag during the pulling tests:

$$\varepsilon = \frac{L - L_0}{L_0} \tag{3.13}$$

We note the experimental fact that the volumetric strain of a soilbag is less than 1% before failure. The volume of a soilbag can thus be considered constant during deformation. Subsequently, the strain $\varepsilon$ of the bag can be related to $\varepsilon_y$ (the component of $\varepsilon$ at the short axis of the soilbag, see Figure 3.43) by:

$$\text{Vol} = B_0 H_0 = BH \tag{3.14}$$

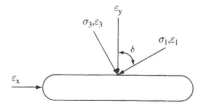

Figure 3.43 Illustration of the inclined stresses acting on a soilbag and the resultant strains.

in which Vol is the volume of the soilbag per unit length (area). Substituting Eq. (3.8) into (3.9), the following equation can be obtained:

$$\sigma_1 = \sigma_3 f(\varepsilon_1) + \frac{2T}{B}\left\{\frac{B}{H}f(\varepsilon_1) - 1\right\} \tag{3.15}$$

If the tensile force $T$ is taken as the value at failure and $f(\varepsilon_1)$ is considered as $K_p$, Eq. (3.15) is thus reduced to Eq. (3.1). Substituting Eqs (3.11), (3.12), (3.13) and (3.14) into Eq. (3.15) yields:

$$\sigma_1 = \frac{f(\varepsilon_1)}{B_0}\left[\sigma_3 B_0 - 2k\varepsilon_y \frac{n+\varepsilon_y-1}{(n+1)(1-\varepsilon_y)}\left\{\frac{1-\varepsilon_y}{f(\varepsilon_1)} - \frac{n}{(1-\varepsilon_y)}\right\}\right] \tag{3.16}$$

The above equation may be used to predict the major principal strain $\varepsilon_1$ of soilbags when the major principal stress $\sigma_1$ along the short axis of the soilbags $(\delta = 0)$ is known, and vice versa.

### 3.6.2  Deformation estimation when soilbags are subjected to major principal stress with an inclination to the short axis of the soilbags $(\delta \neq 0)$

The geometric relationship of the principal stresses and the principal strains is illustrated in Figure 3.43. Based on the assumption that soilbags remain at constant volumes during deformation $(\varepsilon_v = \varepsilon_1 + \varepsilon_3 = 0)$, the following equation may be deduced through the Mohr's strain diagram:

$$\varepsilon_y = \varepsilon_1 \cos 2\delta \tag{3.17}$$

Laboratory tests (Matsuoka et al., 2002a) also showed that the apparent cohesion $c(\delta)$ is a function of the inclination $\delta$ in the form of $c(\delta) = c(0)\cos 2\delta$. As $c(\delta)$ is proportional to the tensile force $T$ along the bag, the coefficient $k$ in Eq. (3.12) may be expressed as:

$$k(\delta) = k(0)\cos 2\delta \tag{3.18}$$

Substituting Eqs (3.17) and (3.18) into Eq. (3.16) yields:

$$\sigma_1 = \frac{f(\varepsilon_1)}{B_0}\left[\sigma_3 B_0 - 2k(\delta)\varepsilon_y \frac{n + \dfrac{\varepsilon_y}{\cos 2\delta} - 1}{(n+1)\left(1 - \dfrac{\varepsilon_y}{\cos 2\delta}\right)}\right.$$
$$\left.\left\{\left(1 - \frac{\varepsilon_y}{\cos 2\delta}\right)\frac{1}{f(\varepsilon_1)} - \frac{n}{\left(1 - \dfrac{\varepsilon_y}{\cos 2\delta}\right)}\right\}\right] \tag{3.19}$$

Associated with Eq. (3.17), the above equation may be used to predict the deformation (the major principal strain $\varepsilon_1$) of soilbags when the major principal stress $\sigma_1$ is inclined to the short axis of the soilbags ($\delta \neq 0$), and *vice versa*.

The stress–strain relationships of soilbags calculated through Eq. (3.19) are plotted in Figure 3.44. The parameters of $n = 5$, $\phi = 25°$, $k = 320\,\text{N/cm}$, $B_0 = 5\,\text{cm}$, $H_0 = 1$ cm and $a = -0.683$ are used in the calculations (Matsuoka *et al.*, 2002a). The biaxial compression tests are conducted on aluminum rod assemblies piled up at various inclinations ($\delta = 0$–$45°$). The test results are recorded in Figures 3.44(a–d). In addition, the unconfined compression test results on soilbags filled with various materials are exhibited in Figure 3.45, where the calculations using Eq. (3.19) with the parameters of $n = 4$, $\phi = 40°$, $k = 450\,\text{N/cm}$, $\sigma_3 = 0$, $B_0 = 40\,\text{cm}$, $H_0 = 10\,\text{cm}$ and $a = -0.278$ are also illustrated. In comparison to the calculated and the experimental results, it can be seen that Eq. (3.19) properly predicts the stress–strain behavior of the soilbags. Assuming the constant volumes of soilbags during deformation, the minor principal strain $\varepsilon_3 = -\varepsilon_1$. Therefore, $\varepsilon_3$ can be obtained as well.

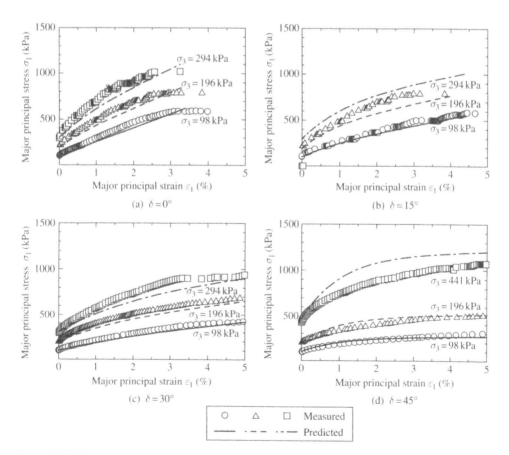

Figure 3.44 The measured and the predicted stress–strain relationships of soilbags at (a) $\delta = 0°$; (b) $\delta = 15°$; (c) $\delta = 30°$; (d) $\delta = 45°$.

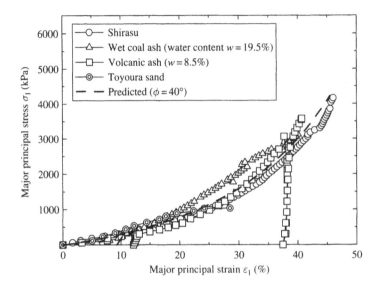

*Figure 3.45* Results of unconfined compression tests on soilbags filled with various materials.

Also, compaction with vibration or preloading may reduce the primary deformation of soilbags. Subsequently, the tensile force $T$ along the bags will develop quickly. The resultant $\sigma_1 \sim \varepsilon_1$ relationship may be patterned as those shown in Figure 3.45. To estimate this effect, we may use the unloading–reloading relationships between the principal stress and the principal strain of the materials inside soilbags (the straight dashed line in Figure 3.42) to acquire the value of $f(\varepsilon_1)$. The deformation (the major principal strain $\varepsilon_1$) of soilbags can thereby be estimated through Eq. (3.19).

In Figures 3.44(a–d), the predicted stress–strain relationships are convex upwards because of the contribution of the confining stress $\sigma_3$. However, when the major principal strain $\varepsilon_1$ increases to large values, the tensile force $T$ developing along the bag will dominate and contribute significantly to the major principal stress $\sigma_1$ in comparison to the effect of $\sigma_3$. Under this circumstance, the stress–strain relationship becomes concave, as shown in Figure 3.45.

The deformation of soilbags can also be directly estimated through the unconfined compression tests. The major principal strain $\varepsilon_1$ measured from this kind of test is somewhat conservative. This is because, in practice, the confining pressure $\sigma_3$ is usually not equal to zero. It should be mentioned that the unconfined compression tests should be carried out on the same soilbags as those used in the field and the soilbags should be compacted in advance before loading so as to approximate field conditions.

We thereby summarize the test results on soilbags filled with different materials, as shown in Figure 3.45, from which the deformation coefficient of soilbags can be acquired. The eventual deformation of soilbags can thus be estimated.

## 3.7 Friction between soilbags

The friction along the contact surfaces between bags, the friction between soilbags, and the friction between soilbags and other materials are experimentally investigated through a series of laboratory tests. The first group of tests is conducted on bags made of PE. As listed in Table 3.3, the friction angles between the bags, $\phi_s$ and $\phi_{pl}$, are taken from the measured shear force–horizontal displacement curves. $\phi_s$ corresponds to the value when two bags commence relative movement, and $\phi_{pl}$ is taken at the peak frictional resistance. The average value of $\phi_s = 15°$, corresponding to the friction coefficient of $\tan \phi_s = 0.27$. The averaged $\phi_{pl} = 23°$, related to the friction coefficient $\tan \phi_{pl} = 0.42$. The relative displacement at the peak frictional resistance is about 5 mm.

The second group of tests is carried out on two soilbags that are piled up vertically, as shown in Figure 3.46. These soilbags are filled with either No. 6 silica sand or Toyoura sand. The maximum and mean grain sizes of the No. 6

*Table 3.3* Test results of the interface friction between bags and between soilbags filled with fine and coarse materials

| Friction conditions | Material inside soilbags | Average frictional angle $\phi_s$ at the onset of sliding | Average frictional angle $\phi_{pl}$ at peak | Friction angle $\phi$ of the material inside soilbags |
|---|---|---|---|---|
| Interface of bags | | 15° | 23° | |
| Interface between soilbags with fine material inside | No. 6 Silica sand | 15° | 23° | 40° |
| | Toyoura sand | 15° | 22° | 40° |
| Interface between soilbags with coarse material inside | Crushed stone | Not measured | 31° | 44° |
| | Rockfill material | Not measured | 43° | 45° |

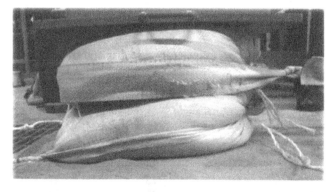

*Figure 3.46* Friction tests between two vertically piled soilbags. The soilbags are filled with fine sand, crushed stone and rockfill materials, respectively.

silica sand are 0.9 and 0.25 mm, respectively. The maximum and mean grain sizes of the Toyoura sand are 0.8 and 0.18 mm, respectively. The internal friction angle of these two kinds of sands is around 40°. The relative displacement at the peak frictional resistance is about 5–10 mm. The friction between the interfaces of these two soilbags is measured, as shown in Table 3.3. For both the No. 6 silica sand and the Toyoura sand, the average values are $\phi_s = 15°$ and $\phi_{p1} = 22°-23°$, respectively. These values approximate the friction angles between the bags. This is because both the No. 6 silica sand and the Toyoura sand are so fine that they do not contribute to the friction along the contact surfaces of the bags.

Furthermore, the third group of friction tests is conducted on soilbags filled with rockfill materials and coarse crushed stones, respectively. The soilbag arrangement is the same as in the Figure 3.46. For the rockfill materials, the maximum and mean grain sizes are 53 and 12 mm, respectively, and the internal friction angle $\phi = 45°$. During the tests, as the rockfill materials are out of the soilbag surface due to the large size of the grain and the angularity, the measured average value at the peak friction resistance $\phi_{p1} = 43°$ approaches the internal friction angle $\phi = 45°$ of the rockfill materials. For the crushed stones, the grains partially come up to the soilbag surface, but not as thoroughly as the rockfill materials. Therefore, it may be seen that, for the crushed stones, the measured $\phi_{p1} = 31°$ is smaller than its internal friction angle $\phi = 44°$ but larger than the friction angle between bags (23°). The test results are listed in Table 3.3.

In the fourth group of tests, two soilbags are connected mouth-to-mouth and placed on the ground forming the base, as shown in Figure 3.47. The third soilbag is located between the two connected soilbags. The inclination of the soilbag base to the horizontal level is represented by $\theta$, and the apparent friction angle between the soilbag base and the third soilbag at the peak horizontal resistance is denoted as $\phi_{p2}$. The friction angle between the third soilbag and the soilbag base is recorded in Table 3.4. It is interesting to see that, on average, $\phi_{p2}$ is almost equal to the sum of $\theta$ and $\phi_{p1}$ although different materials are filled in the bags. The experimental results may be interpreted as follows. A wedge is sitting on a slope surface. The inclination of the slope is $\theta$. The friction angle between the wedge and the slope surface is equal to $\phi_\mu$. If the wedge is pulled horizontally with a force $T$, the

*Table 3.4* Test results of the interface friction between one soilbag and the valley of two soilbags

| Material inside soilbags | Average frictional angle $\phi_s$ at the onset of sliding | Average frictional angle $\phi_{p2}$ at peak | Inclined angle $\theta$ of the valley slope | $\phi_{p2} = \phi_{p1} + \theta$ |
|---|---|---|---|---|
| No. 6 silica sand | 15° | 55° | 30° | 53° |
| Toyoura sand | 17° | 47° | 25° | 47° |
| Crushed stone | Not measured | 55° | 23° | 54° |
| Rockfill materials | Not measured | 61° | 25° | 68° |

*Figure 3.47* Friction tests: one soilbag is located above two soilbags connected mouth-to-mouth.

*Figure 3.48* Friction tests: soilbags filled with rockfill materials are placed above the same type of rockfill materials.

wedge experiences a vertical component $N$. The ratio of $T$ to $N$ is approximately equal to the tangent of the summation of $\theta$ and $\phi_\mu$.

We carry out the fifth group of tests so as to investigate the method of increasing the friction between soilbags and the ground. Rockfill materials are accumulated on the ground as the base. The soilbags filled with the same kind of materials are arranged on the rockfill-material base in a steel-latticed frame, as shown in Figure 3.48. In this case, the friction angle $\phi_{pl}$ between the soilbags and the rockfill materials is almost equal to the internal friction angle of the rockfill materials $\phi = 45°$, owing to the stability of the base. If the steel frame is removed, slight movement can be observed in the rockfill materials and the resultant $\phi_{pl} = 37°$.

In order to further increase the horizontal resistance during sliding, a short pile/concrete block is placed between soilbags and the rockfill materials, as shown in Figure 3.49. In this case, the apparent friction angles $\phi_{pl}$ between the soilbags and the rockfill materials may increase up to $67°$–$72°$. In practical applications, this technique may be employed to stablize the soilbag-constructed retaining walls.

# Chapter 4

# Design approaches of the Solpack method

Soilbags have been used in the construction of embankments and retaining walls, as well as to reinforce soft building foundations. In this chapter, we introduce the design approaches of the Solpack method – constructing structures with soilbags under static and cyclic loadings (Matsuoka *et al.*, 2002b, 2004a). To reach sufficient stability, the applications of the flat soilbags are emphasized herein. The earthquake loading is assessed using the seismic coefficient method with the values of $k_v = 0$ and $k_h = 0.15$. This is equivalent to the inclination of the major principal stress of $8.5°$ to the vertical direction, which is often implemented in the design for civil engineering structures.

## 4.1  Embankment constructed with soilbags

A schematic view of an embankment constructed with soilbags is shown in Figure 4.1. For the safety of the embankment, two essential factors should be considered:

1  whether or not the soilbags will break under the dead weight of the overlaying soilbags and the overburden $\gamma z$; and
2  the determination of the potential failure surface of the embankment.

Regarding the first factor, we use Eq. (3.1) to calculate the bearing capacity of a soilbag. Considering the typical size of a soilbag as $40\,cm \times 40\,cm \times 10\,cm$ (width $B \times$ length $L \times$ height $H$), the tensile strength of a bag per unit length $T = 12\,kN/m$, and the internal friction angle of the materials inside the bag $\phi = 30°$. Under the condition of $\sigma_{3f} = 0$, Eq. (3.1) gives the ultimate bearing capacity of the soilbag $\sigma_{1f} = 660\,kPa$. This is equivalent to the stress caused by a $36.7\,m$ high embankment if the density of the soilbag approximates that of the soils with $\gamma = 18\,kN/m^3$.

As far as the potential failure surface is concerned, as shown in Figure 4.1, we assume a straight line representing the failure surface for simplicity. Accounting

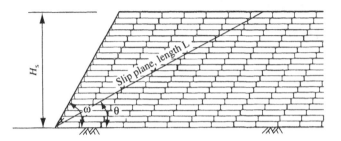

*Figure 4.1* Schematic view of the embankment constructed with soilbags.

for the seismic load, the safety factor $F_s$ along the straight line can be computed by Eq. (4.1)

$$F_s = \frac{(1-k_v) - k_h \tan\left(45° + \frac{\phi}{2} - \delta\right)}{(1-k_v)\tan\left(45° + \frac{\phi}{2} - \delta\right) + k_h} \tan\phi$$

$$+ \frac{\ell \cdot c\,(\delta = 0°)\cos 2\delta}{\left\{(1-k_v)\sin\left(45° + \frac{\phi}{2} - \delta\right) + k_h\cos\left(45° + \frac{\phi}{2} - \delta\right)\right\} W} \qquad (4.1)$$

in which $\delta$ is the angle between the major principal stress and the short axis of the soilbag, $W$ is the weight of the soilbags overlying the sliding line, $l$ is the length of the sliding line and $c(\delta = 0)$ is the apparent cohesion of the soilbag when $\delta = 0°$. At regular service, $k_h = k_v = 0$ are taken when the factor of safety $F_s$ is computed using Eq. (4.1).

Under regular service and seismic load conditions, the variations of $F_s$ with respect to the height $H_s$ of the embankment are respectively illustrated in Figures 4.2 and 4.3. For conservative considerations, we use a smaller internal friction angle $\phi = 30°$ in the calculation for the materials in the soilbags. Furthermore, as the apparent cohesion $c(\delta)$ of soilbags decreases with the increasing inclination $\delta$ of the major principal stress, the largest value possibly encountered in engineering practices $\delta = 15°$ is taken in the calculation of $F_s$. Also, as the seismic load is equivalent to an inclination of 8.5° between the major principal stress and the vertical direction, the equivalent $\delta$ should thus be taken to be larger than the sum of 8.5° and 15° so as to cover the seismic effect. In the present calculation, $\delta = 30°$ is used. It may be seen from Figures 4.2 and 4.3 that the embankment built with soilbags can reach up to 60 m high even if it is constructed under the worst conditions: vertical construction and subject to seismic load. In common engineering practices, an embankment is generally less than 20 m in height (usually 2–5 m). The breakage of soilbags and the sliding failure within soilbags might not occur simultaneously for such an embankment. Moreover, sufficiently large friction between soilbag interfaces ($\phi_{pl} = 23°$; $\tan 23° = 0.42$) would prevent the soilbags from sliding out in the front row caused by the seismic load when the horizontal seismic coefficient $k_h = 0.15$ exists.

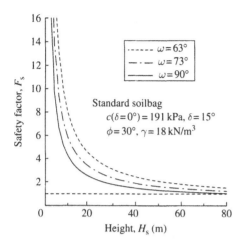

Figure 4.2 The calculated safety factors against the height of embankment in usual circumstances.

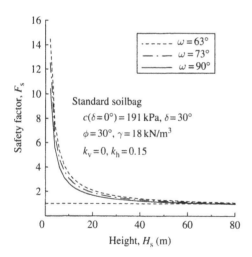

Figure 4.3 The calculated safety factors against the height of embankment under seismic load.

## 4.2 Reinforcement of soft ground with soilbags

Soilbags contribute to the enlargement and deepening of the footing when they are used in the reinforcement of soft foundations, as illustrated in Figure 4.4. This method would eventually increase the bearing capacity of the soft foundations. The probability of the breakage of a single soilbag may be estimated using Eq. (3.1). After the foundation has been reinforced with soilbags, the bearing capacity of the foundations may be computed with the same formula (e.g. Taizagi's formula for

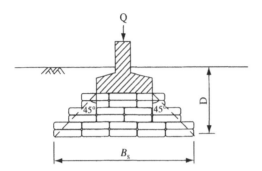

Figure 4.4 Schematic view of the building foundation reinforced with soilbags.

bearing capacity calculation) as used for the foundations without any reinforcement. The ultimate bearing capacity $q_f$ is usually expressed as follows

$$q_f = cN_c + \gamma DN_q + \gamma B_s N_\gamma / 2 \qquad (4.2)$$

in which $N_c$, $N_q$ and $N_\gamma$ are the coefficients of the bearing capacity. However, it is noteworthy that the enlarged width $B_s$ and the deepened depth $D$ of the enlarged footing should be employed. The maximum value of the enlarged inclination for the footing width is assumed to be 45°, as illustrated in Figure 4.4, based on the fact that the apparent cohesion $c(\delta)$ may reduce to zero when the inclination $\delta$ of the major principal stress exceeds 45° (Matsuoka *et al.*, 2002a). If the soilbags beneath the footing do not break under the action of the upper loading, sliding failure among soilbags is almost impossible.

## 4.3  Retaining walls built with soilbags

When subjected to lateral active earth pressure, the major principal stress would incline to the vertical direction (the short axis of a soilbag). For a retaining wall built with soilbags, the inclination between the major principal stress and the vertical direction is denoted as $\delta$ as illustrated in Figure 4.5. The value of $\delta$ is estimated through the following simple case.

Considering that the internal friction angle of the materials inside soilbags $\phi = 30°$, the friction angle along the interface between soilbags and the backfill $\phi_w = 0$, the inclination of the surface of the backfill ground $\beta = 0$, and the horizontal and vertical seismic coefficients are $k_h = k_v = 0$, the inclination $\delta$ may be calculated by:

$$\tan 2\delta = \frac{2\tau_{xy}}{\sigma_x - \sigma_y} = 2\frac{(\frac{1}{2}K_a\gamma z^2)/B_s}{\gamma z - K_a\gamma z} = \frac{z}{2B_s} \qquad (4.3)$$

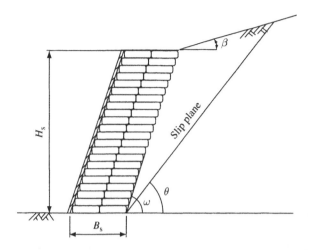

*Figure 4.5* Schematic view of the retaining wall built with soilbags that must be connected mutually in the horizontal direction.

in which $B_s$ and $z$ are the width and the depth of the retaining wall, respectively. Based on this estimation, we have

$\delta = 0$   at   $z = 0$;

$\delta = 22.5°$   at   $z = 2B_s$;   and

$\delta = 28.2°$   at   $z = 3B_s$,

all of which are within the range of $0 \leq \delta < 45°$. Thus, there exists an apparent cohesion in the expression of

$$c(\delta) = c(\delta = 0) \cdot \cos 2\delta \qquad (4.4)$$

This is one of the reasons that retaining walls constructed with soilbags can be nearly vertical.

It is necessary to determine the possible sliding failure surfaces among soilbags or between the lowest level of soilbags and the ground. Due to the lateral active earth pressure, the sliding force is $(1/2)K_a \gamma z^2$ at the depth $z$, in which $K_a$ is the coefficient of the active earth pressure. The frictional resistance force between soilbags is $\gamma B_s z \cdot \tan \phi_{p1}$. The comparison between these two forces suggests that the most probable sliding failure surface among soilbags take place at a large value of $z$. Considering the soilbags filled with fine materials like sands and assuming no breaks of the soilbags, in the equilibrium state (the factor of safety $F_s = 1.0$), we have

$$(1/2)K_a \gamma z^2 = \gamma B_s z \cdot \tan \phi_{p1} \qquad (4.5)$$

Therefore,

$$B_s/z = K_a/(2 \tan \phi_{pl}) \tag{4.6}$$

If the friction angle between the interfaces of bags is $\phi_{pl} = 23°$, Eq. (4.6) reduces to $B_s/z = 0.39$.

Various methods may be used to reduce the critical value of $B_s/z$. Recalling the series of laboratory tests in Section 3.7, we may recognize that soilbags filled with crushed stones or large-grained coarse materials could increase the interface friction angle $\phi_{pl}$ between soilbags up to $30°-40°$. Also, soilbags may be connected mouth-to-mouth and/or placed above the valley of two connected soilbags so that the values of $\phi_{pl}$ or $\phi_{p2}$ increase (Table 3.4). In addition, increasing the interface friction angle $\phi_g$ between the lowest level of soilbags and the ground may also contribute to the stabilization of the retaining walls.

Taking into account these factors, we calculate the stability of a soilbag retaining wall against sliding and overturning using a similar method as for a concrete retaining wall. Various parameters are used to cover the possibility of failure that would lead to a minimum safety factor $F_s$. As illustrated in Figure 4.6, we assume that the inclination of the surface of the backfill ground behind the soilbag retaining wall $\beta = 15°$, and the horizontal and vertical seismic coefficients are $k_h = 0.15$ and $k_v = 0$, respectively. The materials filled in the soilbags have the unit weight $\gamma = 18 \, \text{kN/m}^3$ and the internal friction angle $\phi = 30°$ without cohesion $(c = 0)$. The friction angle between the lowest level of soilbags and the ground $\phi_g = 30°$, and the friction angle between the soilbags and the backfill materials $\phi_w = 20°$. For a soilbag retaining wall, the most probable failure situation is sliding along the contact surface between the lowest level of soilbags and the ground. The minimum safety factor $F_s$ occurs under a seismic load. It should be

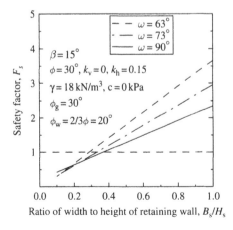

*Figure 4.6* Minimum safety factor of resisting sliding for the soilbag-built retaining wall under seismic load.

noted that, in practical applications, even a small value of cohesion from the backfill materials (e.g. $c = 10\,\mathrm{kPa}$) would benefit significantly the stability of a soilbag retaining wall.

Additionally, Figure 4.6 shows that when the ratio of the width to the height $B_s/H_s$ is between 0.3 and 0.4, the safety factor of a soilbag retaining wall is around 1.0. This corresponds to $B_s/H_s = 0.3$–0.6 for a concrete retaining wall. Therefore, the retaining walls constructed with soilbags are comparable to concrete retaining walls if soilbags are mutually connected.

# Chapter 5

# Applications of the Solpack method

## 5.1 Railway ballast foundations

Ballast is usually accumulated on the surface of a railway foundation. The confining stress $\sigma'$ of the ballast due to self-weight is thus very small, which leads to a small value of shear strength $\tau_f$ ($\tau_f = \sigma' \tan \phi$). This is why, in soft railway foundations, lateral movement of the ballast and settlement of the sleepers are significant. This will do harm to the railways. However, if the railway foundation is reinforced with soilbags that are filled with ballast, tensile forces may be created along the bags when the soilbags are under the loading of trains. Subsequently, the shear strength of the ballast foundation increases greatly, which eventually leads to an increase in the bearing capacity of the railway foundation. The settlement of the sleepers can thus be reduced. The effectiveness of this method has been verified not only through a series of laboratory experiments using aluminum rods and No. 6 crushed stones, but also prototype tests on the ballast foundations in the JR General Research Institute (Kachi *et al.*, 1997; Matsuoka *et al.*, 1998; Matsuoka and Liu, 1999; Matsuoka *et al.*, 2004d). Moreover, this method has been applied in a local JR where the ballast foundation was so weak that the mud of the foundation was compressed under the vibration caused by trains, and harmful settlement had taken place.

### 5.1.1 Verification through laboratory experiments

#### 5.1.1.1 Bearing capacity tests on aluminum rods

In order to study the ballast foundation under railway sleepers, the bearing capacity test is conducted on assemblies of aluminum rods as shown in Figure 5.1. An aluminum block, 12 cm wide by 8.7 cm high, is used to simulate the railway sleeper. The size of the sleeper model is half the size of the real No. 3 sleeper used in Japan. The ballast foundation is simulated with the assemblies of aluminum rods (mixed 5 and 9 mm diameter rods with a weight ratio of 3:2). The mixed assemblies are constructed to 21.2 cm (12.5 + 8.7 cm) in height. The height of the foundation model is half that of the real ballast foundation in Japan. The roadbed beneath the foundation is simulated with an assembly of aluminum rods with mixed diameters of 1.6 and 3 mm, and with a mixing weight ratio of 3:2. In this experimental setup,

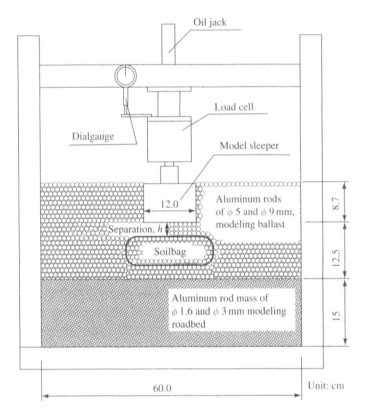

*Figure 5.1* Schematic view of the bearing capacity test on assemblies of aluminum rods to simulate the ballast foundation.

the aluminum block and all the aluminum rods adopted are 5 cm in length. This assembly of aluminum rods is wrapped with a piece of paper to construct a soilbag model 20 cm in width and 6.25 cm in height. The soilbag model is placed under the sleeper. It is noted that dynamic tampers are commonly used as maintenance for ballast foundations in Japan. In the current test, the separation, $h$, between the soilbag and the sleeper is left for the dynamic tampers.

A series of tests are carried out by exerting static load on the sleeper through an oil jack. The following cases are considered:

a    with/without soilbags to reinforce the foundation;
b    under different separation $h = 0, 1, 2, 3$ and 4 cm between the soilbag and the sleeper.

The relationship between the applied static load and the settlement of the sleeper is shown in Figure 5.2. It may be seen that, when $h = 2$–3 cm (about 3–4 times the average size of the aluminum rods), the bearing capacity of the foundation is improved significantly and the sleeper has relatively small values of settlement. In

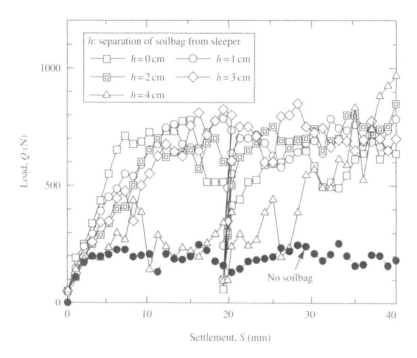

*Figure 5.2* Relationships between the applied static load and the settlement of the assembly of aluminum rods.

these cases, the aluminum rods between the soilbag and the sleeper contact tightly. However, when $h = 4$ cm, the aluminum rods between the soilbag and the sleeper escape laterally, leading to less improvement of the bearing capacity and large values of settlement of the sleeper.

The dynamic bearing capacity tests have also been performed on assemblies of aluminum rods at JR Technical Research Institute (Kanzaki *et al.*, 1997). A sine-waved dynamic stress with a frequency of 20 Hz is applied on the sleeper model. The average, maximum and minimum values of the applied dynamic stress are 29.4, 49 and 9.8 kPa, respectively. Figure 5.3 shows the test results with respect to the relationship between the dynamic loading cycles and the settlement of the sleeper. Without any reinforcement to the foundation, the static bearing capacity of the foundation approaches the value of that under an average dynamic stress of 29.4 kPa. After being reinforced by a soilbag under the sleeper, the static bearing capacity of the foundation doubles the value of that under the maximum dynamic stress of 49 kPa. Moreover, after reinforcement, the settlement of the sleeper is only about 10% of that without any reinforcement under dynamic loading. These experiments also agreed that the smallest settlement occurs when $h = 2$ cm.

### 5.1.1.2  Bearing capacity tests on crushed stones

The effectiveness of reinforcing railway ballast foundations with soilbags has also been verified through a series of static and dynamic bearing capacity tests on

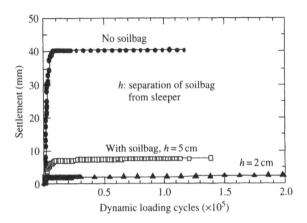

*Figure 5.3* Results of the bearing capacity tests on aluminum rods under dynamic loading (after Kanzaki *et al.*, 1997).

No. 6 crushed stones (Matsuoka *et al.*, 1998). Figure 5.4 shows a schematic view of the test set-up. The No. 6 crushed stones have maximum and average grain sizes of 20 and 8 mm, respectively. River sand with maximum and average grain sizes of 2.5 and 0.6 mm is used to simulate the roadbed. The sleeper, 60 cm long by 12 cm wide by 8.5 cm high, is made of wood. The soilbag, 20 cm wide by 20 cm long by 7.5 cm high, is constructed by filling a geogrid bag with No. 6 crushed stones (mesh span: 1.5 cm, tensile strength: 88.2–98 kN/m). The relationship between the applied static load and the settlement of the sleeper is recorded in Figure 5.5. It is obvious that the bearing capacity of the No. 6 crushed stones increases after soilbag

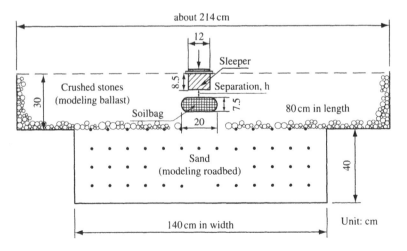

*Figure 5.4* Plan view of the bearing capacity tests on No. 6 crushed stones.

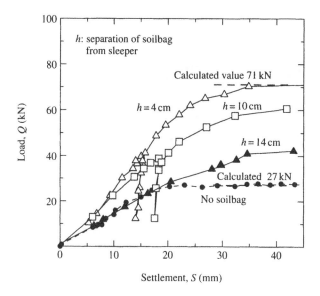

*Figure 5.5* Relationship of the applied load and the settlement from the static bearing capacity tests on No. 6 crushed stones.

reinforcement. For the current crushed stones ($D_{50} = 8$ mm, $D_{max} = 20$ mm), the maximum possible separation between the soilbag and the sleeper is 4 cm (about five times the $D_{50}$). Under the same value of separation ($h = 4$ cm), 27,000 cycles of dynamic loading (relevant vibration frequency of 30 Hz, vibration force of 2.94 kN and static force of 49 kN) are applied. Since the average grain size of the ballast commonly used in JR is $D_{50} = 3$–4 cm, we can calculate from these laboratory tests that the separation between the soilbag and the sleeper in fields is around 15–20 cm. This value of separation is large enough for the dynamic tampers in the traditional maintenance for ballast foundations. Nevertheless, we still propose that the separation should not be larger than 10–15 cm (about three to four times $D_{50}$) in practical applications, for safety reasons.

### 5.1.1.3 Prototype tests on real ballast foundations

The JR Technical Research Institute has performed a series of dynamic prototype bearing capacity tests on real ballasts (Kachi *et al.*, 1997). A schematic view of the experimental set-up is shown in Figure 5.6. The ballasts used in the test have maximum and average grain sizes of 63.5 and 42 mm, respectively. The soilbags are constructed by filling high-strength Vectron bags with the ballasts and are, 35 cm long by 40 cm wide by 15 cm high. The large sleeper of 210 cm long by 30 cm wide by 14 cm high is used. Five soilbags are placed directly on to well-compacted ground. Ten centimeter thick ballasts are placed between the soilbags and the sleeper separation. Two steel rails with a length of 60 cm are placed on the sleeper. A sine-waved dynamic load with a

(a) Experimental set-up of the prototype test on real ballasts

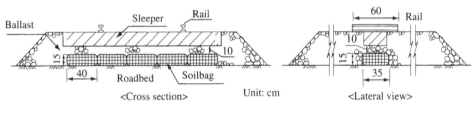

(b) Plan views of the experimental set-up

*Figure 5.6* Prototype bearing capacity tests on real ballasts conducted by the JR Technical Research Institute (after Kachi et al., 1997).

vibration frequency of 20 Hz is applied on the rails. The maximum, minimum and average of the applied dynamic load are 313.6, 19.6 and 166.6 kN, respectively. The settlement of the sleeper is recorded in Figure 5.7(a) under the 100,000 cycles of the dynamic load. It is seen that, when $h = 10$ cm and 100,000 cycles of dynamic loading is applied, the settlement of the sleeper with the reinforcement of soilbags is only 1/2 of that without any reinforcement. Moreover, the initial gradient of the settlement curve with reinforcement is smaller than that without any reinforcement. Therefore, we may conclude that the influence of the cyclic dynamic load to the eventual settlement of the sleeper is smaller when the ballast foundation has been reinforced with soilbags.

Under unloading–reloading cycles, the gradient of the settlement curve is usually considered to be equivalent to the elastic coefficient of railway structures. It is directly related to the oscillation of trains and thereby affects the satisfaction of train passengers. Considering this regard, we continued the static loading tests are continued to carry out after the 100,000 cycles of the dynamic loading. The relationship between the applied load and the settlement of the sleeper is shown in Figure 5.7(b). It is seen that, under unloading–reloading cycles, the gradients

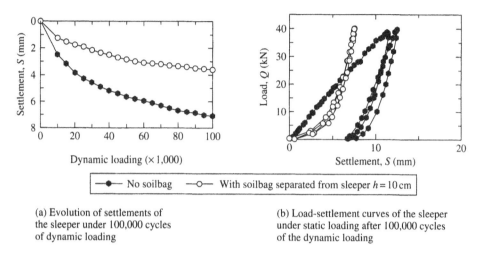

(a) Evolution of settlements of
the sleeper under 100,000 cycles
of dynamic loading

(b) Load-settlement curves of the sleeper
under static loading after 100,000 cycles
of the dynamic loading

*Figure 5.7* Results of the prototype bearing capacity tests on the real ballasts.

of the curves are similar whether the foundation is reinforced with soilbags or not. In other words, reinforcement with soilbags does not change the elastic coefficients of the railway structures.

### 5.1.2   Applications to a local Japanese Railway

Located at a foothill (Figure 5.8(a)), a local JR suffered from overflows in the rainy season, with large amounts of mud squeezed from the roadbed. The settlement of the sleepers was greater than 20 mm. The routine maintenance of inserting new ballast into the foundation with dynamic tampers had to be done frequently for the proper operation of the railway. The Solpack method is thus applied to the reinforcement of weak railway foundations. A schematic view of reinforcement with soilbags is shown in Figure 5.9. The construction procedures are as follows:

- Remove the sleepers from the rails and excavate all the ballasts (about 42 cm in depth) and the roadbed foundation (10 cm in depth).
- After compacting the excavated foundation with vibrators (called plate-compactors), place a layer of regular soilbags on it. The soilbags are made of PE bags filled with No. 5 crushed stones and are constructed to a size of 40 cm × 40 cm × 10 cm. These soilbags are connected using high strength ropes and compacted thoroughly with vibrators (plate-compactors), as shown in Figure 5.8(b).
- Put a layer of special soilbags above the layer of the regular soilbags (Figure 5.8(c)), above which are the ballasts and the sleepers (Figure 5.9). These

special soilbags are compacted with vibrators. The special soilbags, sized 35 cm × 60 cm × 15 cm, are made of high-strength Vectron bags filled with fresh ballasts.

- Reset the sleepers and insert fresh ballasts between the sleepers and the soilbags (Figure 5.8(d)).
- Compact the ballasts with dynamic tampers.

*Figure 5.8* Railway foundation and its reinforcement using soilbags for a local JR line.

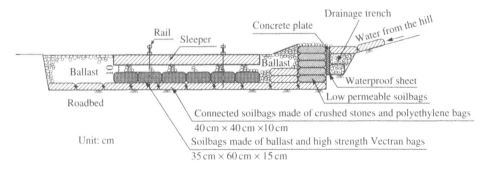

*Figure 5.9* Schematic view of a weak railway foundation reinforced with soilbags.

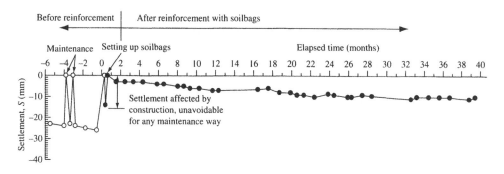

*Figure 5.10* Settlements of the sleepers before and after reinforcement by soilbags.

The construction can be carried out only in the interval after the final train of the day passes and before the first train comes the next day. It took two nights to complete the reinforcement of the 13 m long railway foundation. A snapshot of this process is provided in Figures 5.8(b–d). In addition, in order to mitigate the influence on the railway caused by the mudflow due to rainfall, a small retaining wall was constructed near the foothill with soilbags and waterproof sheets. A drain trench was also built with the above-mentioned special soilbags between the retaining wall and the hill so as to divert the ground water from the hill.

After the reinforcement of the railway foundation, settlement at the rail joints on the sleepers was measured for more than three years. The results are shown in Figure 5.10. Before the railway foundation was reinforced with soilbags, the settlement was up to 20 mm although routine maintenance with dynamic tampers was frequently implemented. After the construction, the settlement was greatly reduced, less than 10 mm within more than three years. Moreover, the average oscillation acceleration of the trains decreased from $0.12\,g$ ($g$: gravity acceleration) to $0.04g$. The effectiveness of this method has been demonstrated.

## 5.2   Soft foundation reinforcement

With appropriate arrangements, soilbags have been applied to reinforce soft foundations of low-level buildings (1–3 floors). The designs depend on the strength of the foundations. To reinforce weak/soft building foundations, the general principle is to arrange the soilbags under the footing of the building as wide (widening effect) and as deep (penetration effect) as possible so as to disperse the load from the upper building. Figure 5.11 is one application where soilbags are placed under raft foundation of a building, corresponding to different $N$-values. It is worth mentioning that different layers of soilbags should be overlapped rather than vertically piled up along the force dispersive direction. Also, two layers of soilbags are placed under the whole raft foundation. In order to effectively withstand the external forces under large deformation, the soilbags under the footings should preferably be connected mouth-to-mouth and be

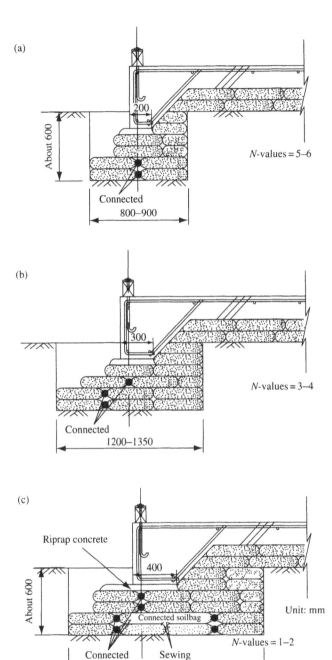

(a)

About 600

200

Connected
800–900

*N*-values = 5–6

(b)

300

Connected
1200–1350

*N*-values = 3–4

(c)

Riprap concrete

About 600

400

Connected soilbag

Unit: mm

*N*-values = 1–2

Connected  Sewing
1600–1800

*Figure 5.11* Design plan for soft foundation reinforcement using soilbags (lower-level buildings, raft foundation).

sewed end-to-end, as illustrated in Figure 5.12. Soilbags may constitute a flexible structure together with the footing of buildings and the adjacent soft foundation, as described in Figure 5.11, due to their interactions and the varying stiffnesses. The flexible deformation of soilbags can also be observed from Figure 5.13. The space between soilbags is filled with either crushed stones/excavated soils or small soilbags filled with 1/5–1/4 the amount of material. Each layer of soilbags is fully compacted with vibrators (plate-compactors). A soilbag usually has a height of 8–10 cm. Various kinds of materials like crushed stone, sand, concrete waste, asphalt waste, tile waste and excavated soil may be used in siolbags.

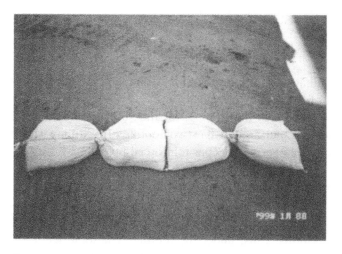

*Figure 5.12* Method of four-soilbag connection.

*Figure 5.13* Flexible deformation of soilbag assembly against external force under a significantly large settlement of the footing model (soilbags are connected horizontally).

To elucidate the effectiveness of this reinforcement method, we estimate the bearing capacity of the soft ground with soilbag arrangement as shown in Figure 5.11(c). Since soilbags have very high strength (more than 1.4 MPa), as indicated in Section 3.1, we may assume that soilbags are unlikely to break/crack under the usual loading from the upper building. Moreover, under vertical load, the apparent cohesion of soilbags of 270 kPa may be obtained using Eq. (3.3) if the soilbag parameters $T = 12$ kN/m, $\phi = 44°$, $B = 40$ cm and $H = 10$ cm are taken. As large apparent cohesion is induced due to the tensile force of the bags, the most probable sliding failure may take place in the foundation beneath the soilbags rather than inside the assembly of soilbags. Regarding the design shown in Figure 5.11(c), the contributions of soilbags are summarized as follows:

- The third-lowest soilbags beneath the footing take effect firstly under the upper building load;
- The fourth-lowest soilbags and those around the footing come to be effective when the settlement of the footing develops;
- When the settlement continues developing, all the soilbags including those under the raft foundation and around the two footings (the figure shown is symmetrical) comprise a larger and deeper reinforced body.

Here, we take only the third-lowest soilbags beneath the footing into account. The contributions of soilbags are regarded as widening and deepening the footing. The effective width of the soilbag assembly is assumed to be 120 cm, the total width of the third-lowest soilbags just under the footing. The total height of the four layers of soilbags beneath the footing is 40 cm, the extremely soft ground is considered with the soil properties of $\gamma_t = 16$ kN/m$^3$, $c = 10$ kPa and $\phi = 0°$. As shown in Figure 5.11(c), the width of the footing is 40 cm and the depth between the footing and the ground surface is 20 cm. Under these conditions, the bearing load of the foundation per unit length of the footing is calculated as 21.84 kN/m before reinforcement. After reinforcement with soilbags, the corresponding value is 73.2 kN/m. In other words, the bearing load increases 3.4 times. The effectiveness of this reinforcement method for soft foundations is thereby demonstrated.

Various applications of this reinforcement method using soilbags in different cities in Japan are introduced below.

### 5.2.1  In YC cho, Ibaraki Prefecture

The first application using soilbags to reinforce soft building foundations is described in the previous section, as shown in Figure 5.11(c). The design details and the construction process of soilbags have also been presented in Figures 3.28, 3.33 and 3.34. For a single-storey house foundation with an area of 20 m × 10 m, about 4,500 soilbags are needed for reinforcement. The material in the soilbags is crushed stone, which is cheap and easily attainable near the construction. In the five years since the completion of this reinforcement, the foundation and the house have remained in very good condition. Moreover, the reduction of vibration is an additional benefit.

### 5.2.2  In FS-cho, Ibaraki Prefecture

Being a wetland often partially covered with water, the site in FS-cho, Ibaraki Prefecture is extremely weak. The ground soil has an $N$-value of 1–2. Nevertheless, this area is to be built upon. If piles were to be used to reinforce the foundation, the driving depths would have to have to reach more than 30 m. As an alternative, we plan to use soilbags to reinforce this soft foundation. The site has an area of about 20 m × 10 m; about 6,000 soilbags are needed for such weak ground. The material used to form the soilbags is crushed stone. The arrangement of soilbags is similar to that shown in Figure 5.11 (corresponding to the case with $N$-values of 1–2), and the construction details are illustrated in Figure 5.14. After reinforcement with soilbags, the house built on the reinforced foundation has been problem-free. However, the septic tank has been observed to have a significant inclination due to the settlement of the ground without reinforcement. This phenomenon demonstrates the effectiveness of reinforcement using soilbags. Moreover, it is worth noting that the traffic-induced vibration has been reduced in this field as well.

### 5.2.3  In MB city, Chiba Prefecture

There was a weak foundation in MB city, Chiba Prefecture. The ground had been inundated by heavy rainfalls. The waterlogged ground was so weak that crushed stones would sink if they were directly placed on the ground. The $N$-value of the ground soil was around 2. In the reinforcement plan, P bags were filled with crushed stones to construct soilbags. The arrangement of about 6,500 soilbags is similar to that shown in Figure 5.11 (corresponding to

(a)                                                    (b)

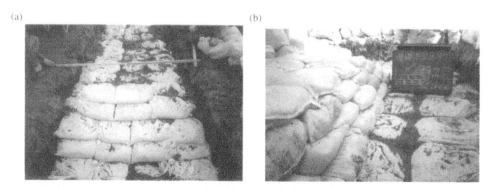

*Figure 5.14* Reinforcement of a soft building foundation in FS-cho, Japan.

(a)                                    (b)

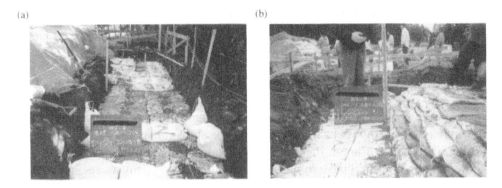

*Figure 5.15* Reinforcement of a soft building foundation in MB city, Japan, where soilbags are constructed in water.

the case with $N$-values of 1–2) but with modifications. The soilbags below and surrounding the septic tank were fastened onto the septic tank. In this case, the septic tank would not float. Figure 5.15 shows the construction process. Five years have passed since the reinforcement. The ground has been stable and there have been no signs of settlement.

### 5.2.4  In KR-cho, Chiba Prefecture

A construction site in KR-cho, Chiba Prefecture has a high ground water level, only 50 cm beneath the ground surface. Two-storey wooden houses are to be constructed on this water-rich weak ground. In order to avoid excavation in such a high groundwater-level foundation, soilbags are used to reinforce the wet and soft ground. Four soilbags are connected as a group and placed under the footing, as shown in Figure 5.16. Three layers of soilbags are laid directly on to the soft ground. Also, soilbags are constructed under the raft foundation. After the arrangement, thorough compaction is carried out on the soilbags. A piece of plastic sheet is then placed

(a)                                    (b)

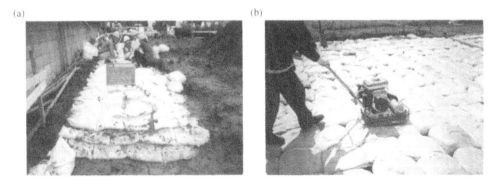

*Figure 5.16* Reinforcement of a soft building foundation in KR-cho, Japan, where the groundwater level is high, only 50 cm beneath the ground surface.

above the soilbags for waterproofing. The steel-reinforced concrete is cast on the plastic sheet. Since the construction, settlement observation has been conducted for a couple of years. No sign of settlement has been observed. This field case demonstrates that the soilbag reinforcement method can also be used in water-rich foundations.

### 5.2.5  In S city, Miyagi Prefecture

The site in S city, Miyagi Prefecture, has been flooded to the extent that a bucket can be floated on the water, as shown in Figure 5.17. If a person were to stand in this wetland, his feet would be buried as deep as 30 cm. This site is set aside for the construction of single-storey houses. The following are commonly used methods for handling such types of wetland:

- Giving up this site for building construction;
- Using a number of piles to reinforce the soft foundation, but the cost is soaring;
- Using chemical agents to drain off the water.

If piles were driven into the ground, the neighbors would suffer excessive noise during the long construction process. If chemical stabilization agents were used, there would be environmental pollution since nearby areas include agricultural land and community districts. Our experience of handling wetland inspires us using soil-bags to reinforce this waterlogged foundation. The soilbags are filled with crushed stone. The design plan is illustrated in Figure 5.18, in which four soilbags are connected mouth-to-mouth and sewed end-to-end. These soilbags and two other unconnected soilbags are placed on the ground below the underground water level. Three layers of soilbags with the same formation are constructed. Furthermore,

(a)                                         (b)

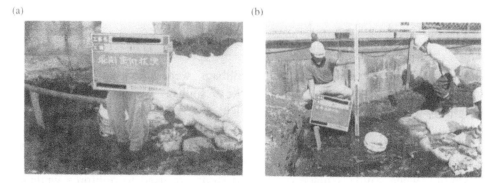

Figure 5.17 Waterlogged foundation in S city, Japan, where soilbags are being constructed.

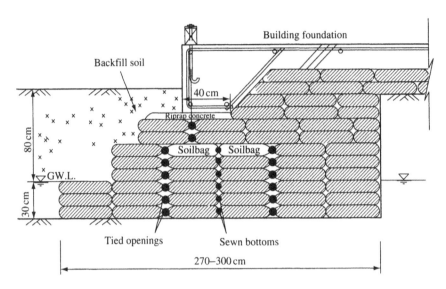

*Figure 5.18* Design plan of waterlogged building foundation reinforced with soilbags in S city, Japan.

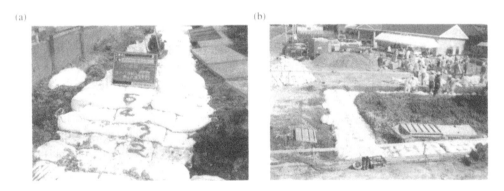

*Figure 5.19* Reinforcement of waterlogged weak building foundation with soilbags in S city, Japan.

another five layers of the connected soilbags were piled up, as shown in Figures 5.18 and 5.19. In total eight layers of soilbags were constructed on the waterlogged ground. After this reinforcement, the ground was able to withstand heavy construction machines, for example a backhoe. Five years have passed since the reinforcement. The houses have been working very well and no sign of settlement has been seen. It may be seen from this case that, besides high strength and flexibility, when soilbags made of crushed stone are used to reinforce wetland, they also demonstrate good drainage ability.

### 5.2.6  In TK city, Hokkaido

The description of the site and the soilbag construction process have been carried out in previous sections. The effect of vibration reduction has also been confirmed by observing the fluctuation of the water surface in three glass cups simultaneously when a JR freight train passed. Here we note that the soilbags we used for reinforcement are filled with sandy soil and gravel obtained from the nearby seashore. The construction process of soilbags is shown in Figure 5.20. Soilbags composed of gravels are expected to have positive effects on frost heave prevention in cold areas like Hokkaido (northern Japan).

### 5.2.7  In OT city, Hokkaido

On this site, two purposes dominate the use of soilbags: one is to increase the bearing capacity of the soft ground and the other is to prevent frost heave. The material used in the soilbags is coarse gravel. The soilbags are connected mouth-to-mouth and sewed end-to-end. Four soilbags are fastened together as one group. A number of soilbags with this special connection method are placed under the footing of the building. The plan view of the soilbags under the footing and the whole steel-reinforced concrete raft foundation is illustrated in Figure 5.21. In order to investigate the frost heave in this soilbag-reinforced foundation, we divide the foundation into two parts. One part of the foundation is wrapped with Styrofoam boards. The temperature changes of the foundation are monitored during the whole winter. The measurements of the temperature show no difference between the

(a)

(b)

*Figure 5.20* Reinforcement of a soft building foundation with soilbags in TK city, Japan.

(a)                                          (b)

*Figure 5.21* Reinforcement of a soft building foundation with soilbags in OT city, Japan.

two parts of the foundation. This case shows that the soilbags filled with gravel are effective in frost heave prevention in cold regions, which is due to the large air voids in gravel that may prevent the decrease in temperature and the rise of capillary water.

### 5.2.8 In ST city, Osaka

The construction site is located 3 m away from a Grade A river. The riverbank is covered with blocks/stones. A three-storey building of steel-framed structure was to be constructed in this field. The load from the upper building would spread into the riverbank. It was required that the load should be transmitted downward directly at least 2–3 m deep before dispersing laterally to protect the riverbank. To handle this problem, the usual method is to drive piles or cast a concrete retaining wall as the foundation of the buildings. Based on our understanding and experience on soilbags, we proposed a plan for the foundation reinforcement as shown in Figure 5.22. The asphalt wastes were filled into PE bags to construct soilbags. Parallel to the riverbank, a trench 1 m wide, 2 m deep and 13 m long was dug. The soilbags were connected mouth-to-mouth and piled up inside the trench layer by layer. Before arranging the soilbags as the foundation, a layer of geogrid was placed around the circumference of the trench, as shown in Figure 5.23. Six years have passed since completing the reinforcement, the building and the nearby area have been working normally. This application demonstrates that the Solpack method is environmentally friendly, using recycled materials to form soilbags. In addition, the construction cost is much cheaper than some of the commonly used methods, such as pile foundation or concrete retaining walls. Also, the construction process is time-effective and less noisy. Moreover, the Solpack method can also be used as the building-support bodies, functioning like piles and concrete retaining walls (Matsuoka, 1999).

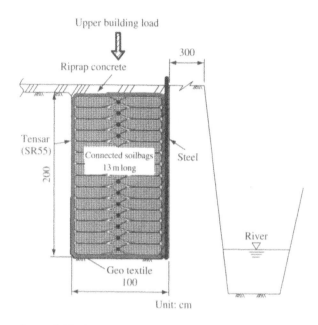

*Figure 5.22* Cross-section of the reinforced foundation using soilbag piles in ST city, Japan.

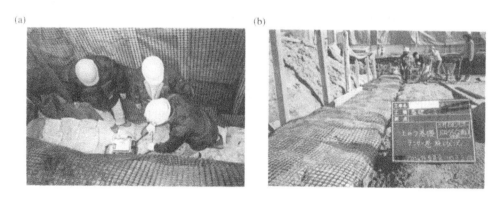

*Figure 5.23* Construction of the soilbag-pile foundation in ST city, Japan.

### 5.2.9   Reinforcement for elevator foundation in K city, Kyoto

When an elevator is located on a soft foundation, vibration can be easily felt when the elevator starts moving or stops suddenly. Soilbags are used to reinforce the foundation. As shown in Figure 5.24, five layers of soilbags are placed on the ground and thoroughly compacted. After the reinforcement, the elevator moves smoothly and safely. Little vibration is felt. Thus, soilbags are also effective in this type of earth reinforcement.

*Figure 5.24* In K city, Japan, a soft foundation where an elevator was located is reinforced with five layers of thoroughly compacted soilbags.

## 5.3 Soilbag piles

Soilbag piles can be constructed by piling up soilbags vertically with the flat plane parallel to the ground. Soilbag piles are largely applied to reinforce shallow soft building foundations (a few meters in depth). The principles of the soilbag piles and some field applications are introduced below.

### 5.3.1 Principle of soilbag piles

The concept of the soilbag piles is introduced in Figure 5.25, in which the soilbag piles are constructed under the footing of buildings. Buried into the soft ground with

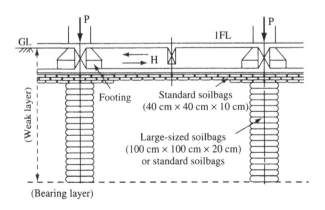

*Figure 5.25* Schematic view of the foundation constructed with soilbag piles.

shallow depth, the soilbag piles essentially function as a piled raft foundation. As summarized in Chapter 3, soilbags are environmentally friendly, having almost the same unit weight as soil foundations and flexible deformation. Also, if soilbags are protected from ultraviolet rays like sunlight, they are durable and can be used for more than 50 years. Moreover, soilbags have very high compressive strength when subjected to external forces along their short axis. Subsequently, we may expect that soilbag piles may work as the raft foundation and reduce the settlement of the soft foundation under heavy load from buildings. Using this reinforcement method, two important factors should be considered during the construction process. One is to fill the space between soilbags and the surrounding area. The other is to thoroughly compact the soilbags. Therefore, if soilbags are embedded in the ground with the flat plane parallel to the ground surface, the above two conditions are satisfied. Soilbag piles may be considered as an ideal reinforcement method for soft foundation.

Furthermore, the flat plane of soilbags is large. Piled vertically as a soilbag pile, the stiffness of the soilbag piles is more adapted to the surrounding area in comparison to the concrete or steel piles. Besides, soilbag piles have an equivalent damping coefficient as high as 0.1–0.3 (the same as that of rubber). The materials inside soilbags and the friction between soilbag interfaces may behave as a damping agent, separating seismic shear. Therefore, from a dynamic analysis point of view, soilbag piles are effectively resistant to horizontal forces caused by earthquakes, acting as the seismic isolation base (Yamamoto *et al.*, 2002).

### 5.3.2   Applications of soilbag piles

By September 2002, soilbag piles had been successfully applied to reinforce building foundations in ten cases in Japan. We select two representative cases to introduce as follows.

#### 5.3.2.1   In TY city, Chiba Prefecture

Figure 5.26 shows the distribution of the $N$-values along the depth of the foundation in a construction site in TY city, Chiba Prefecture. As it is a reclaimed land of rice field, the top 3 m of soil is very weak. We propose using soilbag piles to reinforce the soft foundation. Six soilbag piles are distributed in the circumference of the site, as shown in Figure 5.27. During the construction, the intervals between soilbags and the surrounding area are filled with the same material as is filled in the soilbags. Full compaction on soilbags and the material in the intervals are also important, as demonstrated in Figures 5.28 and 5.29.

#### 5.3.2.2   In SG city, Kanagawa Prefecture

There is a reclaimed piece of land which was formerly rice field in SG city, Kanagawa Prefecture. The strength of the top 2.5 m of soil is very weak, as in

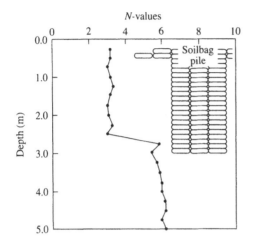

Figure 5.26 Distribution of N-values in the soft building foundation in TY city, Japan.

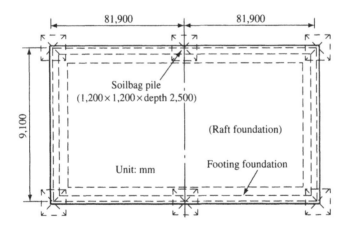

Figure 5.27 Arrangement of the soilbag piles to reinforce a soft building foundation in TY city, Japan.

to the previous case. A two-storey building was to be constructed in this site, as shown in Figure 5.30. The building was to be 30 m long. A foundation beam was added under the second floor. Considering the building configuration, we arranged 16 soilbag piles under the building foundation, as illustrated in Figure 5.30. The commencement and the completion of the construction are shown in Figures 5.31 and 5.32. It should be noted that the intervals between soilbags and the surrounding ground were thoroughly filled. Also, the soilbags were fully compacted. Soilbags were connected in order to increase stability of the soilbag piles.

*Figure 5.28* Construction of soilbag piles (completion of 3 layers) in TY city, Japan.

*Figure 5.29* Compaction of soilbags during the construction in TY city, Japan.

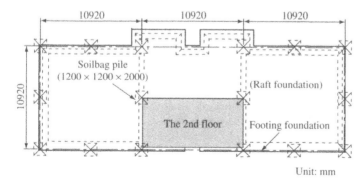

*Figure 5.30* Arrangement of the soilbag piles to reinforce a soft building foundation in SG city, Japan.

*Figure 5.31* Construction of soilbag piles in the site at SG city, Japan (the first layer).

*Figure 5.32* Completion of the soilbag piles in the site at SG city, Japan.

## 5.4 Retaining wall

### 5.4.1 Restoration of a sliding slope in Fukuoka Prefecture

A slope collapsed in Fukuoka Prefecture, Japan. The slope was 10 m in height and constructed almost vertically, as shown in Figure 5.33(a). If soil were simply backfilled to restore the slope, the slope would be unstable. If concrete were cast

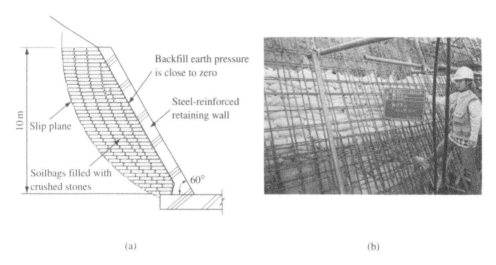

Figure 5.33 Soilbags are used to construct a retaining wall to stabilize a collapsed slope in Fukuoka Prefecture, Japan.

on the collapsed slope surface, it would be very expensive due to the large size of the collapsed slope. Based on our understanding of and experience with soilbags, we proposed a method of constructing a thin steel-reinforced concrete retaining wall with soilbags as the backfill (Matsuoka, 1999). The soilbags were filled with crushed stone. Each soilbag was constructed to a height of 9 cm for stability. About 8,000 to 10,000 soilbags were used. It only took three weeks to complete the entire restoration work. During the construction, the piled soilbags stood so stable that the workers even suggested removing the thin steel-reinforced concrete retaining wall. Nevertheless, the thin steel-reinforced concrete retaining wall was still constructed in order to shut out the ultraviolet rays from the PE-made soilbags. Soilbags filled with crushed stones are ideal as the backfill for the retaining wall to encourage drainage.

In order to investigate why the assembly of soilbags was so stable, the assembly of aluminum rods was wrapped to simulate the effect of soilbags. These models were piled up as shown in Figure 5.34. Unconfined compression tests were conducted. Twenty layers and three columns of soilbag models were piled up vertically. A vertical load was applied to them. The soilbag model was composed of aluminum rods with dimensions of 4 cm × 1 cm × 5 cm (width × height × length). It was surprising to find that the assemblies of the soilbag models were very stable, even without horizontal support. There was no sign of failure under the maximum load of the device used. This experiment helps us to understand that soilbags may be used as an emergency measure. Figure 5.33 shows the effectiveness of the applications to slope stability, earth pressure reduction and so on.

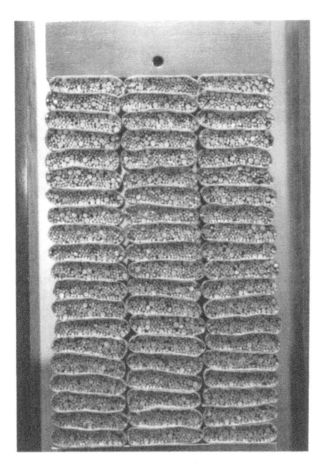

*Figure 5.34* Unconfined compression tests on assemblies of model soilbags.

### 5.4.2  Retaining wall in NO city, Aichi Prefecture

*In situ* direct shear tests were used to acquire the shear strength of the ground soils. Near the bottom of the retaining wall foundation, soil samples were excavated with a backhoe. The tests were not successful because the groundwater was only about 25 cm below the ground surface, as shown in Figure 5.35(a). The excavated ground soil was so weak that a 1 m long steel rod was easily pushed into the ground by hand. We stopped excavating the root foundation, but started to pile up soilbags directly on the original ground surface. As shown in Figure 5.35(b), the retaining wall was built with 20 layers of soilbags (Matsuoka *et al.*, 2000c). Each layer consisted of four soilbags connected mouth-to-mouth with ropes and sewing end-to-end. The connected soilbags were pulled to produce a tensile force along the bags quickly. The material inside the soilbags was crushed stone, which is cheap and easily acquired near the construction site. The intervals between soilbags were also filled with the same type of crushed stone. The

(a) Waterlogged ground

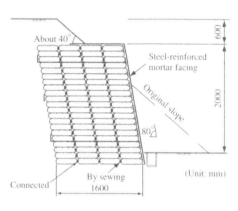

(b) Arrangement of the connected soilbags

(c) Using soilbags to construct a retaining wall

(d) Completion of the soilbag retaining wall.

*Figure 5.35* A soilbag constructed retaining wall in NO city, Japan (height 2 m, total length 50 m, slope angle 80°).

soilbags were fully compacted with vibrators (plate-compactors) layer by layer during construction. The completed retaining wall, with a height of 2 m and total length of 50 m, inclined 80° with respect to the horizontal direction. About 5,500 soilbags of 40 cm × 40 cm × 10 cm (width × length × height) were used at this site. Figure 5.35(c) shows the completion of the piled soilbags. In order to protect soilbags against ultraviolet rays, the surface of the retaining wall was covered with a thin layer of steel-reinforced mortars, as shown in Figure 5.35(d). The area in front of the retaining wall was used as a parking lot.

### 5.4.3 Retaining wall in MS city, Shizuoka Prefecture

Soilbags were used to construct a retaining wall on a Kanto loam slope. As shown in Figure 5.36(a), the upper ground was a parking lot. In order to enlarge the parking lot,

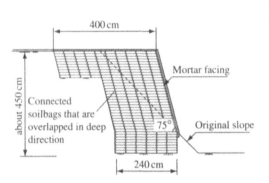

(a) Cross-section of the soilbag retaining wall

(b) Front view of the soilbag retaining wall

(c) Down view of the soilbag retaining wall

(d) Completion of the soilbag retaining wall

*Figure 5.36* A retaining wall is constructed on a Kanto Loam slope by piling up soilbags in MS city, Japan, which is 4.5 m in height with a slope angle of 75° and a total length of about 21 m.

a retaining wall was constructed with soilbags in front of the parking lot. The base of the soilbag retaining wall was constructed vertically, while the portion above the boundary line of the neighboring ground was constructed as steep as 75°. The retaining wall was 4.5 m in height and about 21 m in total length. In the lower part of the retaining wall, each layer consisted of six connected soilbags, whilst in the upper part of the retaining wall, each layer was arranged with ten connected soilbags. In order to increase the stability of the retaining wall, soilbags were overlapped layer by layer in the longitudinal direction along the retaining wall. The soilbags were filled with construction waste containing tile, which was free of charge. The space between soilbags was filled with the same construction waste. Compaction was performed thoroughly on soilbags layer by layer during construction. Figure 5.36(b) shows the completion of the soilbag retaining wall, which looks like a stone-cast retaining wall. After the completion of the soilbag retaining wall, some soilbags on the crest of the retaining wall had to be removed since drainage pipes needed to be arranged (Figure 5.36(c)). It was very difficult to remove the soilbags to leave the space to the drainage pipes because all these soilbags had been connected and thoroughly compacted, although it was done eventually.

During the construction, the transmission capacity of vibration through soilbags was also investigated by striking the surface of the retaining wall with a backhoe bucket. At the start of the striking, the vibration could be felt. But the vibration lasted only for a very short period, demonstrating the effectiveness of the vibration reduction of soilbags. In the similar way as shown in Figure 5.35(d), a thin layer of steel-reinforced mortars was cast on the outside surface of the retaining wall to protect the soilbags against ultraviolet rays (Figure 5.36(d)). In this project, about 30,000 soilbags were used.

### 5.4.4  Restoration works in Miyake Island, Japan

In 2000, a large volcanic eruption took place in Miyake Island, Japan. In the restoration works, the Tokyo Metropolitan decided to use soilbags to construct two retaining walls and a diversion embankment in order to protect houses and main roads from the mudflows caused by the volcanic eruption. Before the full-scale construction, a trial retaining wall with soilbags 5 m high and 6 m long was built. As shown in Figure 5.37, the collapsed mountain slopes were modeled with

*Figure 5.37* A trial retaining wall is built with soilbags (5 m high and 6 m long) in Miyake Island, Japan.

several steel-made covering platens. On both the left and right sides of the covering platens, the large-size soilbags were piled up to construct the retaining wall model. The large soilbag was 1 m wide, 1 m long and 20–25 cm high, whilst the standard soilbag was 40 cm wide, 40 cm long and 8–10 cm high. On the right side of the covering platens (the opposite side of the photograph), a piece of sunlight-proof sheet was used to cover the surface of the soilbag retaining wall temporarily, taken as an emergency countermeasure. On the left side of the covering platens (the front side of the photograph), an L-shaped thin concrete wall was constructed to cover the soilbags and cut off sunlight to protect the soilbags against ultraviolet rays. In addition, this wall also stabilized the soilbags, acting as a permanent structure. The L-shaped concrete thin wall is 1 m high and 2 m long.

After the success test of the trial retaining wall with soilbags, two full-scale retaining walls with a slope of 73° were constructed in Miyake Island, as shown in Figures 5.38 and 5.39. Also, a diversion embankment was constructed with approximately 10,000 large soilbags (approx. 1 m × 1 m × 0.2 m). With a length of 436 m and height of 2–5 m, the diversion embankment was used to direct the mudflows along the road by the mountain in the Tsubota district. The soilbags were filled with scoria, the volcanic debris that needed to be cleared. As shown in Figure 5.40, four soilbags were piled up as one unit with a height of 0.8 m, and in total five unit layers of soilbags were constructed as the banks of the diversion embankment. In order to increase the stability of the banks against external lateral forces, every

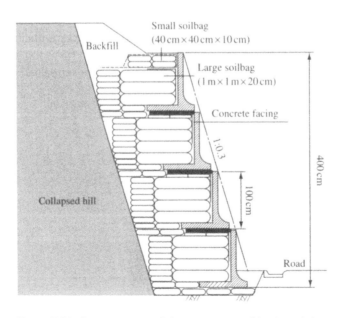

*Figure 5.38* Cross-section of the retaining wall built with large soilbags in Miyake Island, Japan.

*Figure 5.39* Construction of the retaining wall in Miyake Island, Japan.

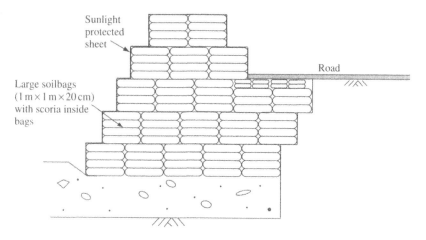

*Figure 5.40* Typical cross-section of debris-diversion embankment constructed with large-sized soilbags in Miyake Island.

unit layer of soilbags was wrapped with geogrid. A piece of sunlight-proof sheeting was placed on the outside surface of the banks so as to protect soilbags against ultraviolet rays. As shown in Figure 5.41, it is a magnificent and graceful structure, which can effectively absorb the tremendous energy of the debris flows with its

(a)                                           (b)

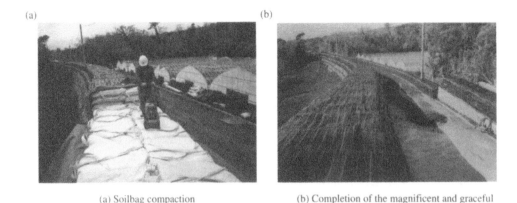

(a) Soilbag compaction              (b) Completion of the magnificent and graceful
                                     debris-diversion embankment

*Figure 5.41* Construction of the debris-diversion embankment with large-size soilbags as part of the restoration works in Miyake Island.

pliancy. Moreover, unlike concrete structures, the banks constructed with soilbags are easy to remove and recover on site (Matsuoka and Liu, 2003).

## 5.5 Tunnel lining

Consider how a tunnel can be excavated under a 3,000 m high mountain and why concrete linings can support the overburden load of $\gamma z (= 2\,\text{tf}/\text{m}^3 \times 3,000\,\text{m} = 6,000\,\text{tf}/\text{m}^2 \approx 60,000\,\text{kN}/\text{m}^2)$. This is because the greatest part of the overburden has been transmitted to the surrounding area of the excavated tunnel due to the arching. Figure 5.42 shows a trap-door test on the rods of photo-elastic material, in which the arch-shaped transmission of interparticle forces is illustrated

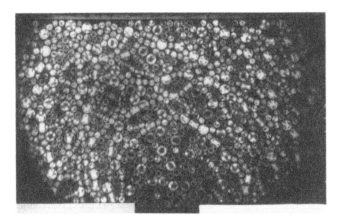

*Figure 5.42* The trap-door test on the photo-elastic material showing the arch-shaped transmission of interparticle forces.

when the bottom plate descends (Matsuoka, 1999). The arch-shaped transmission lines of interparticle forces are formed around the spacious area caused by the descending of the bottom plate. The formation of the arch-shaped lines of the major compressive principal stresses is called the arching action. Due to the arching action, the upper load hardly transmits to the bottom plate but transmits instead to the surrounding area of tunneling. As excavating a tunnel inevitably disturbs the surrounding area, the soils/rocks in these areas are loose. Due to the arching action, the areas surrounding the tunnel receive most of the overburden loads whilst the tunnel only bears a lesser percentage. As a result, it is possible to excavate tunnels only with steel or concrete lining.

Steel or concrete tunnel lining is a kind of arching structure. The stone-piled arching structure is strong enough to resist the compressive forces from the upper overburden. Meanwhile, as described in Section 3.1, soilbags also have very high compressive strength. Due to their characteristics, we have successfully constructed arch structures with soilbags. It should be pointed out that the arch structures built with soilbags could not be used as the lining for the usual tunnels excavated in mountains. Instead, the soilbag tunnels may be used when the arch-shaped structure is built in advance, and the embankment constructed above.

### 5.5.1  Loading tests on an arch structure model constructed with soilbags

In order to have a better understanding of the arch action and the failure mechanism of the arch structure built with soilbags, we used aluminum rods to construct an arch structure as a 2D soilbag model, as shown in Figure 5.43(a) (Matsuoka *et al.*, 2000e). The 5 cm long aluminum rods with diameters of 1.6 and 3 mm were mixed at a ratio of 3:2 by weight. The internal friction angle of the aluminum rods is $\phi = 25°$. In Figure 5.43(b), the model is 60 cm in width and 30 cm in height. An H-shaped steel plate with a cross-section of 10 cm × 10 cm was placed on the surface of the model, on which a load was applied through an oil cylinder. The soilbag model was constructed with the same type of aluminum rods and sized at $B = 3$ cm in width and $H = 1.2$ cm in height. Due to the limit of the loading capacity of our loading device, one piece of extremely weak paper (with a tensile strength of 12.8 N/cm) was used to wrap the aluminum rods to form one soilbag model.

A schematic view of the test is shown in Figure 5.43(b). The arch structure was constructed to $2R = 10$ cm as the inner width, $2(R + B) = 16$ cm the outer width, 7.5 cm the inner height and 10.5 cm the outer height. The construction of the entire model is as follows:

1   an arch structure made with thick paper was built and placed on the ground;
2   the soilbag models (the wrapped aluminum rods) were arranged along the surface of the paper arch structure;
3   the same type of aluminum rods (discrete) were placed on the test box to form the shown model.

(a) Arching structure model constructed with aluminum rods

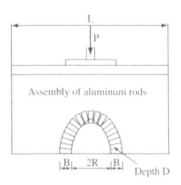

(b) Schematic view of the arching structure

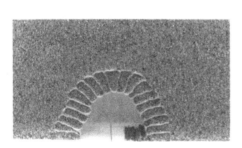

(c) Failure of soilbags at the abutment of the arching structure

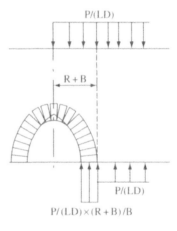

(d) Stresses acting on the abutment of the arching structure

*Figure 5.43* Tests on an arching structure model constructed with aluminum rods.

The test began by gradually applying a load from the upper plate through an oil jack until some soilbag models failed due to breakage of the wrapping paper. The failure of the soilbag models took place at the two abutments of the arch structure, as shown in Figure 5.43(c). In order to predict the value of the load at failure, some assumptions were made. As illustrated in Figure 5.43(d):

1   the load applied on the upper plate was considered as an equivalent uniformly distributed surcharge on the upper plate $P/(LD)$;
2   in the right side area of the dashed line, the ground was assumed to be receiving the same amount of uniformly distributed surcharge from the upper plate;
3   in the area between the central line and the dashed line, the ground above the abutment was assumed to be receiving the same amount of uniformly distributed surcharge from the upper plate.

In this case, each abutment of the arch structure withstands a stress of

$$\frac{P}{LD} \times \frac{R+B}{B}$$

whilst the ground on the right side of the dashed line receives a stress of $P/(LD)$. Here $L = 60\,\text{cm}$, $D = 5\,\text{cm}$, and the weight of the aluminum rods is ignored.

We considered the stresses acting on the soilbags at the abutments of the arch structure. As stated in Section 3.1, Eq. (5.1) describes the stresses acting on the soilbags at failure:

$$\sigma_{1f} + \frac{2T}{B} = K_p \left( \sigma_{3f} + \frac{2T}{H} \right) \tag{5.1}$$

where $K_p = (1 + \sin\phi)/(1 - \sin\phi)$, $\sigma_{1f}$ and $\sigma_{3f}$ are the major and minor principal stresses, respectively, acting on the long and short axis of the soilbag model, $T$ is the tensile force per unit width of the bag and $\phi$ is the internal friction angle of the material inside the soilbag. Based on the above assumption as illustrated in Figure 5.43(d), the force $N$ acting on the soilbag at the abutment of the arch structure is expressed as

$$N = \frac{P(R+B)}{L} \tag{5.2}$$

At failure, the force $N$ acting on the soilbag ($B$ in width, $H$ in height and $D$ in depth) is equal to $\sigma_{1f}BD$. From Eq. (5.1), we have

$$N = \sigma_{1f}BD = \left\{ K_p \left( \sigma_{3f} + \frac{2T}{H} \right) - \frac{2T}{B} \right\} BD \tag{5.3}$$

If $\sigma_{3f} = 0$ is assumed for the soilbags at the abutment of the arch structure, the applied external upper load $P$ at the breakage of the soilbags at the abutment of the arch structure is predicted as follows:

$$P = \frac{BDL}{(R+B)} \left\{ \frac{1 + \sin\phi}{1 - \sin\phi} \times \frac{2T}{H} - \frac{2T}{B} \right\} \tag{5.4}$$

Using Eq. (5.4), the load is calculated as $P = 4.93\,\text{kN}$ on the conditions that $B = 3\,\text{cm}$, $H = 1.2\,\text{cm}$, $D = 5\,\text{cm}$, $L = 60\,\text{cm}$, $R = 5\,\text{cm}$, $T = 12.8\,\text{N/cm}$ and $\phi = 25°$. On the other hand, the measured values of $P$ in the two tests are equal to $4.58\,\text{kN}$ and $5.13\,\text{N}$, respectively. The average value of these two measurements is $4.86\,\text{N}$, which agrees closely with the predicted value of $4.93\,\text{N}$. This indicates that the above calculation procedure is possible and reasonable. It is surprising to see that the upper load $P$ reaches $4,900\,\text{N}$ when the soilbags break, which is much higher than the tensile load of the 5 cm wide wrapping paper $12.8 \times 5 = 64\,\text{N}$ (Matsuoka *et al.*, 2000e).

### 5.5.2  *Construction of a trial arch structure*

A trial arch structure was constructed with an inner width of 2 m and an inner height of 1.73 m by piling real soilbags (width 40 cm, length 40 cm, height 10 cm), as shown in Figure 5.44. The PE bags with a tensile strength of $T = 117.6\,\text{N/cm} = 11.76\,\text{kN/m}$ were used. Various materials like No. 6 crushed stone, sand, glass slag, etc. were used to fill the soilbags. An arched inner support

(a)

(b)

*Figure 5.44* Construction of an arching structure (inner width $= 2\,\text{m}$) with real soilbags.

of the soilbags was made in advance with table wood, wood and concrete blocks inside the arch. Three rows of soilbags were piled up along the arch. The soilbags on the upper part of the arch structure were connected mouth-to-mouth so as to prevent them from falling due to dead weight. After the arrangement of the soilbags, the inner support was removed.

The soilbag arch structure may be used as culverts, above which embankments or roads may be built. Using Eq. (5.4), we can estimate the possible height of the embankment built on the soilbag arch structure. For example, if the material inside

(a)

(b)

*Figure 5.45* Construction of a trial arching structure with an inner width of 5 m by piling up specially made geotextile soilbags (Kubo *et al.*, 2001).

the soilbags has an internal friction angle of 44°, the soilbags have a density of $17.6\,\text{kN/m}^3$, the soilbags arch structure may support a 38.5 m high embankment.

$$\left.\begin{aligned}\gamma z \times (R+3B) \times D &= 3B \times D \left\{\frac{1+\sin\phi}{1-\sin\phi} \times \frac{2T}{H} - \frac{2T}{B}\right\} \\ 17.6z\,(1+1.2) \times D &= 1.2 \times D \left\{5.55 \times \frac{2 \times 11.76}{0.1} - \frac{2 \times 11.76}{0.4}\right\} \\ \therefore\ z &= 38.5\ \text{m}\end{aligned}\right\} \qquad (5.5)$$

Although this is a rough estimate, the result that a 38.5 m high embankment can be built above the arch structure of soilbags is much more than would previously been imagined; it is also very interesting (Matsuoka et al., 2001).

Inspired by the success of the 2 m wide trial arch structure constructed with soilbags, a 5 m wide arch structure was built with special soilbags made of geo-textiles, which was about the width of half the actual tunnels. Special soilbags were used, which were 150 cm long and had a trapezoid section of 35 cm in the short axis, 45 cm in the long axis and 150 cm in height. The material filled in the special soilbag was crushed stone. At the site, special soilbags were piled with a crane symmetrically on the left and right sides of a steel arch-shaped support that was erected in advance. Finally, a special soilbag with a regular triangular section of 150 cm in each length was put on the top of the arch structure, as shown in Figure 5.45. Another special soilbag was placed on the regular triangular soilbag. After the arrangement of the special soilbags, the inner steel support was lowered slightly and then removed slowly. After the removal of the steel support, about 25 cm of settlement occured at the top of the arch structure immediately. Nevertheless, the arch structure built with soilbags was still very stable, as shown in Figure 5.45(b). It has been standing there for more than one year. If it were subjected to earth pressure from the surrounding embankment, the arch structure of soilbags would be more stable. There are many arch structures made of stones around the world. However, it is probably the first time such a large-sized arch structure with soilbags has been constructed (Kubo et al., 2001).

# Chapter 6

# Natural vegetation planted in soilbags

As we have entered the new millennium, one of the most serious problems we face is aggravation of the earth's environment. If corporations are to enjoy sustainable growth and development, they must look to environmentally friendly products as well as technologies that can help to reduce the burden on the environment. This consciousness is also increasingly seen in some civil engineering projects, including slope engineering. The traditional ways to green slopes are either spreading vegetation seeds on slope surfaces or planting trees in pots. However, the drawbacks of these traditional methods lie in that:

1  it is difficult to carry out on steep slopes;
2  naturalized plants may not harmonize with the local environment; and
3  grass vegetation generally survives for only a short period.

However, if native vegetation cuttings are planted inside soilbags or using soilbags filled with surface soils, the above-said defects may be overcome. In this chapter, we introduce the tests on the growth of native vegetation cuttings in soilbags and the spontaneous germination of vegetation seeds in soilbags filled with surface soils.

## 6.1  Growth of native vegetation cuttings in soilbags

The following twelve kinds of native vegetation cuttings have been planted in soilbags:

1  *Celtis sinensis* var. *japonia*,
2  *Zelkova serrata*,
3  *Cinnamomum camphora*,
4  *Rhoddendron pulchrum*,
5  *Citrus tachibana*,
6  *Phinus × yedoensis* cv. *Yedoensis*,
7  *Quercus myrsinnefolia*,
8  *Lagerstroemia indica*,
9  *Illicinm religiosum*,

10   *Melia azedarach* var. *subtripinnuta*,
11   *Rhaphiolepis indica* var. *umbellate*, and
12   *Mallotus japonicus*.

Each of these native vegetation cuttings has been inserted into soilbags through holes bored with a gimlet. The material filled in the soilbags is river sand with maximum and minimum diameters of grain size of 2.5 and 0.055 mm, respectively. Although the best time to plant vegetation is the spring and rainy seasons, we planted the vegetation cuttings into soilbags in July, August and September, respectively, in order to investigate which kind of vegetation could survive in the worst planting season. The soilbags were watered every day. Figure 6.1 shows the growth of the vegetation planted in soilbags. The survival days of the vegetation planted in September are marked in Figure 6.2. It can be seen that the

*Figure 6.1* Vegetations planted in soilbags.

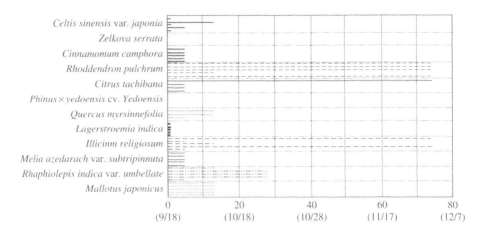

*Figure 6.2* Survival periods of vegetation planted in September.

*Rhoddendron pulchrum* survives for the longest time, while the situations of the *Illicinm religiosum* and the *Rhaphiolepis indica* var. *umbellate* are also good. These three kinds of vegetation are only slightly influenced by the timing of planting recommended in soilbags.

## 6.2  Spontaneous germination of vegetation seeds in soilbags filled with slope mantle soils

If soilbags are filled with slope mantle surface soils, they may be directly used to green slope because the soils contain vegetation seeds (Matsuoka *et al.*, 2004c). Figures 6.3 and 6.4 show the spontaneous germination of vegetation seeds in three

*Figure 6.3* Vegetation in large mesh soilbags (mesh size: 6 mm × 6 mm and 13 mm × 2 mm).

*Figure 6.4* Vegetation in small mesh soilbags (mesh size: 2 mm × 2 mm).

different soilbags with coarse meshes (6 mm × 6 mm and 13 mm × 2 mm) and fine meshes (2 mm × 2 mm), respectively. A thin layer of white cotton sheet is attached to the inner surface of the bags of both coarse and fine meshes. Soilbag filled with a certain amount (about 2 liters) of slope mantle soils is the so-called "green soilbags". Similar to other green soilbags, some tree-typed vegetations germinate in the green soilbags with coarse meshes (Figure 6.3). However, as shown in Figure 6.4, only grass could germinate in the fine mesh soilbags. Tests were carried

*Figure 6.5* Soilbags filled with Masado.

*Figure 6.6* Slope reinforcement with green soilbags.

out on soilbags filled with Masado (soils from completely decomposed granite) as well in order to make a comparison. As Masado does not have vegetation seeds, no vegetation can germinate in these soilbags (Figure 6.5).

It may be more effective to plant native vegetation cuttings in such soilbags that are filled with slope mantle soils. The seeds of vegetation contained in mantle slope soils may germinate and grow rapidly. The native vegetation cuttings may root inside soilbags. The soilbag vegetations contribute to the slope stability.

Figure 6.6 shows an application of green soilbags to reinforce a steep slope under a traffic road. It is seen that the vegetation grows very well inside the soilbags, and is harmonious with the environment.

# Chapter 7

# Concluding remarks

Soils are essentially frictional materials. Wrapping frictional soil materials in a bag is indeed an innovative idea. An interesting phenomenon of soilbags is that after wrapping soils in a bag, frictional materials (gravel, sand and crushed stone, etc.) are transformed to cohesive-frictional materials (with apparent cohesion). This is different from the common reinforcement methods of adding adhesive agents or cements. The apparent cohesion developed due to the tensile force of the bag makes the soilbags behave with high compressive strength. Moreover, soilbags have the ability of converting the external force which is the "enemy" against foundations into the "friends" of the foundations due to the action of tensile forces developed along the bags. More advantages have been demonstrated in field cases, including the amazing bearing capacity, the reduction of traffic-induced vibration resistance to earthquake, the prevention of frost heave and the capability of reinforcing waterlogged soft ground. In addition, the recycling of waste materials, easy construction, and less noise and vibration during construction have made soilbags more privileged than other common earth reinforcement methods. Therefore, reinforcement with soilbags is more effective and reliable than the commonly used horizontal sheet earth reinforcement methods.

Soilbags may be used as permanent materials instead of temporary ones through quality control to guarantee their performance. Quality-controlled soilbags are constructed by wrapping soils in particular bags. The major factors of the quality-controlled soilbags include: usage of high-quality bags, selection of the materials filled in particular bags, specifying the amount of filling materials, optimizing the soilbag arrangement, and compacting the soilbags during construction. The quality of soilbags may be determined using *in situ* tests such as the static or dynamic plate loading tests.

The usual bag materials used for soilbags are PE and PP, which are cheap and whose properties are stable enough to resist acid and alkali. When soilbags are buried into the ground where sunlight is shut out, they would become semi-permanent materials. Since the image of the term "soilbag" is not so attractive, the quality-controlled soilbags would be renamed as "Solpack" (Sol means soil in French), indicating that they are new and can be used as permanent construction materials. Reinforcement using Solpack is called the Solpack method. We sincerely hope that this Solpack method would be widely used in earth reinforcement and civil engineering construction all over the world, including developing countries.

# References

Binquest, J. and Lee, K.L. 1975a. Bearing capacity tests on reinforced earth slabs. *Journal of the Geotechnical Engineering Division*, ASCE, Vol. 101 (GT12), Proc. Paper 11792: 1241–1255.

Binquest, J. and Lee, K.L. 1975b. Bearing capacity analysis of reinforced earth slabs. *Journal of the Geotechnical Engineering Division*, ASCE, Vol. 101 (GT12), Proc. Paper 11793: 1257–1276.

Chen, Y. 1999. Deformation and strength properties of a 2D model soilbag and design method of earth reinforcement by soilbags. *Report to Venture Business Laboratory*, Nagoya Institute of Technology (in Japanese).

Hirao, K., Yasuhara, K., Tanabashi, Y. and Ochiai, H. 1997. Bearing capacity improvement of soft clay reinforced with geogrids. *Journal of Geotechnical Engineering*, JSCE, No. 582/III-41: 35–45 (in Japanese).

Huang, C.C. and Tatsuoka, F. 1990. Bearing capacity of reinforced horizontal sandy ground. *Geotextiles and Geomembranes*, 9: 51–82.

Kachi, T., Miyamoto, H., Matsuoka, H., Tateyama, M. and Kojima, K. 1997. Model tests on improvement method for the bearing capacity of ballast foundation under railway sleepers by soilbags. In: *Proceedings of the 52th Annual Conference of JSCE*, IV-388: 776–777 (in Japanese).

Kanzaki, H., Kachi, T., Matsuoka, H., Tateyama, M. and Kojima, K. 1997. Model tests on improvement method for the bearing capacity of ballast foundation under railway sleepers by soilbags. In: *Proceedings of the 32th Japan National Conference on Geotechnical Engineering*, 1249: 2503–2504 (in Japanese).

Kubo, T. and Yokota, Y., Ito, S., Matsuoka, H. and Liu, S.H. 2001. Trial construction and behaviors of an arching structure with large-sized soilbags. In: *Proceedings of the 36th Japan National Conference on Geotechnical Engineering*, 1065: 2099–2100 (in Japanese).

Matsuoka, H. 1999. *Soil Mechanics*. Morikita Publication Co. Ltd. (in Japanese).

Matsuoka, H. 2002. A new earth reinforcement method with soilbags. *Journal of Geotechnical Engineering*, JSCE, 87: 89–92 (in Japanese).

Matsuoka, H. and Liu, S.H. 1999. Bearing capacity improvement by wrapping a part of foundation. *Journal of Geotechnical Engineering*, JSCE, No. 617/III-46: 235–249 (in Japanese).

Matsuoka, H. and Liu, S.H. 2003. New earth reinforcement method by soilbags ("Donow"). *Soils and Foundations*, 43(6): 173–188.

Matsuoka, H., Takagi, N. and Nishii, M. 1992. An effective reinforcing method for improving bearing capacity of granular foundation. In: *Proceedings of the 47th Annual Conference of JSCE*, III-577: 1194–1195 (in Japanese).

Matsuoka, H., Liu, S.H., Ueda, T. and Nakamura, Z. 1998. Model tests on improvement method for the bearing capacity of crushed stone foundation under railway sleepers by soilbags. In: *Proceedings of the 33th Japan National Conference on Geotechnical Engineering*, 1186: 2377–2378 (in Japanese).

Matsuoka, H., Chen, Y., Kodama, H., Yamaji, Y. and Tanaka, R. 2000a. Mechanical properties of soilbags and unconfined compression tests on model and real soilbags. In: *Proceedings of the 35th Japan National Conference on Geotechnical Engineering*, 544: 1075–1076.

Matsuoka, H., Yamaguchi, K., Maeda, K. and Kodama, H. 2000b. Building vibration reduction by improving foundation using soilbags. In: *Proceedings of the 35th Japan National Conference on Geotechnical Engineering*, 546: 1079–1080 (in Japanese).

Matsuoka, H., Yamaguchi, K., Liu, S.H., Kodama, H. and Yamaji, Y. 2000c. Construction cases of retaining walls and building foundation improvement by soilbags. In: *Proceedings of the 35th Japan National Conference on Geotechnical Engineering*, 545: 1077–1078 (in Japanese).

Matsuoka, H., Liu, S.H., Kodama, H. and Kachi, T. 2000d. Reduction of settlement for ballast foundation under railway sleepers by soilbags. In: *Proceedings of the 35th Japan National Conference on Geotechnical Engineering*, 547: 1081–1082 (in Japanese).

Matsuoka, H., Liu, S.H., Izuka, Y. and Nakamura, J. 2000e. Fundamental study on arch-shaped tunnel lining made of soilbags. In: *Proceedings of 55th Japan National Conference on JSCE*, III-B084 (in Japanese).

Matsuoka, H., Liu, S.H., Kubo, T. and Yokota, Y. 2001. Tunnel lining with an arching structure constructed by soilbags. In: Adachi *et al.* (eds): Proceeding: *Modern Tunneling Science and Technology*, Swets & Zeitlinger, pp. 975–978.

Matsuoka, H., Liu, S.H., Yamamoto, H., Shimao, R., Hasebe, T. and Fujita, T. 2002a. Strength anisotropy of soilbag assembly. In: *Proceedings of the 37th Japan National Conference on Geotechnical Engineering*, 371: 737–738 (in Japanese).

Matsuoka, H., Liu, S.H., Yamamoto, H., Hasebe, T., Shimao, R. and Fujita, T. 2002b. Design method for construction of soilbag assembly. In: *Proceedings of the 37th Japan National Conference on Geotechnical Engineering*, 372: 739–780 (in Japanese).

Matsuoka, H., Liu, S.H., Hasebe, T. and Shimao, R. 2003a. Deformation property of soilbags and its prediction. In: *Proceedings of the 38th Japan National Conference on Geotechnical Engineering*, 377: 753–754 (in Japanese).

Matsuoka, H., Liu, S.H., Shimao, R. and Hasebe, T. 2003b. An environment-friendly earth reinforcement method by soilbags ("Donow"). In: Leung *et al.* (eds): *Proceedings of 12th Asian Regional Conference on Soil Mechanic and Geotechnical Engineering*, pp. 501–504.

Matsuoka, H., Liu, S.H. and Kachi, T. 2003c. Settlement control for railway's roadbed foundation using soilbags. In: Di Benedetto *et al.* (eds): *Proceedings of 3rd International Symposium on Deformation Characteristics of Geomaterials*, LYON, France, pp. 1287–1292.

Matsuoka, H., Liu, S.H., Hasebe, T. and Shimao, R. 2004a. Deformation-strength properties and design methods of soilbag assembly. *Journal of Geotechnical Engineering*, JSCE, No. 764/III-67: 169–181.

Matsuoka, H., Muramatsu, D., Liu, S.H. and Inoue, T. 2004b. Reduction of environment ground vibration by soilbags. *Journal of Geotechnical Engineering*, JSCE, No. 764/III-67: 235–245.

Matsuoka, H., Masuda, M, Asano, Y. and Miwa, O. 2004c. A slope tree planting method by natural vegetation with soilbags. In: *Proceedings of 39th Japan National Conference on Geotechnical Engineering* (in Japanese).

Suzuki, T., Yamashita, S., Matsuoka, H. and Yamaguchi, K. 2000. Effect of wrapped gravel on prevention of frost heaving. In: *Proceedings of the 35th Japan National Conference on Geotechnical Engineering*, 308: 609–610 (in Japanese).

Yamamoto, S. and Matsuoka, H. 1995. Simulation by DEM for compression test on wrapped granular assemblies and bearing capacity improvement by soilbags. In: *Proceedings of the 30th Japan National Conference on SMFE*, pp. 1345–1348 (in Japanese).

Yamamoto, H., Matsuoka, H. and Yamaguchi, K. 2002. Construction cases of column foundations by piling up soilbags. In: *Proceedings of the 37th Japan National Conference on Geotechnical Engineering*, 692: 1377–1378 (in Japanese).

Yamamoto, H., Matsuoka, H., Simao, R., Hasebe, T. and Hattori, M. 2003. Cyclic shear property and damping ratio of soilbag assembly. In: *Proceedings of the 38th Japan National Conference on Geotechnical Engineering*, 379: 757–758 (in Japanese).

Printed and bound by CPI Group (UK) Ltd, Croydon, CR0 4YY

01/11/2024

01782599-0012